REPORTING *the* GREATEST GENERATION

Barrett McGurn

Fulcrum Publishing
Golden, Colorado

Library of Congress Cataloging-in-Publication Data

McGurn, Barrett.
   Yank, the Army weekly : reporting the greatest generation / Barrett McGurn.
      p. cm.
   Includes bibliographical references.
   ISBN 1-55591-296-6 (pbk. : alk. paper) 1. Yank. 2. McGurn, Barrett. 3. World War, 1939-1945—Journalism, Military—United States. 4. World War, 1939-1945—Personal narratives, American. 5. Soldiers—United States—Biography. 6. War correspondents—United States—Biography. I. Title.
   D799.U6M35 2004
   940.54'1273'05—dc22

            2004015483

ISBN 1-55591-296-6

Printed in the United States of America
0 9 8 7 6 5 4 3 2 1

Editorial: Daniel Forrest-Bank, Katie Raymond
Design: Jack Lenzo
Cover design: Patty Maher
Cover image: Mason Pawlak's *YANK* cover photograph of Angaur.
   (*Courtesy of* YANK/ *Photographer Mason Pawlak*)

Fulcrum Publishing
16100 Table Mountain Parkway, Suite 300
Golden, Colorado 80403
(800) 992-2908 • (303) 277-1623
www.fulcrum-books.com

To the Sad Sacks of World War II,
conquerors of tyranny

Contents

Preface

YANK, The Army Weekly, was a phenomenon of World War II, which provided an insight into the nature of the American soldier that always will be valid.

The World War II man in uniform was no superman, no jingoist. He was no different from those who came before or after him. He endured the months of boredom and the moments of terror that are a soldier's fate. He suffered, while at the same time he and his comrades persevered because his country was under attack, just as it was on September 11, 2001, that "nine-one-one" day when the comforting telephone number for emergency help took on a sinister reverse meaning.

YANK was the friend and colleague of the World War II soldier and was itself part of the enlisted U.S. Army. It let off soldier steam with "Mail Call" letters to the editor allowing troops to air their beefs at a global level. With its "What's Your Problem?" section providing frank, if not always the wished-for answers, it was a sort of secular chaplain. With its pinup pictures of gorgeous women back home, *YANK* spoke to the longings of young men who sometimes went months without seeing a woman. With its eyewitness staff reports of courage and death on battlefields around the world, the soldier weekly helped ease the strain of combat by sharing its very horror.

YANK violated all military and journalistic logic. With a circulation of 2,250,000, and a readership double that number created entirely by privates, corporals, and sergeants without officer intervention,

YANK, The Army Weekly

YANK was a major morale instrument and propaganda device not controlled by lieutenants, captains, majors, colonels, or even generals—which indeed flustered some of America's own generals, including the blood-and-guts Patton and the centralist MacArthur.

Caesar provided his own history of a "Gaul divided in three parts" and Britain's commander-in-chief Churchill recorded his particular version of World War II, but it was left to *YANK*'s enlisted men to record America's own week-by-week diary of what the mid-twentieth century global war was like.

YANK was the first periodical to achieve global publication. Faced with how to reach troops on six continents in timely fashion when transportation was scarce, slow, and imperiled, *YANK*'s small staff set up 21 editions in 17 countries; the main magazine was edited in New York and a few pages were substituted locally. Army planes carried the main page forms to regional *YANK* editors and printers in Panama, England, Puerto Rico, Trinidad, Italy, China, Burma, Egypt, Iran, France, Strasbourg, India, Hawaii, Australia, Saipan, Okinawa, the Philippines, and even Japan.

The magazine's staff was part of the Army and had its own small share in the 1945 victory, along with the bombers, the artillery, the infantry, and the warships. Yet it was not the size of a division, nor a regiment, nor even a battalion. It was simply a military company with its single six-stripe first sergeant ruling from inside an office building on Manhattan's East 42nd Street. From there the sergeant's troops, answerable to him, fanned out across the planet to man bureaus, often of a single soldier, in Iceland, Nassau, Greenland, British Guiana, the Alcan Highway, Alaska, the Fiji Islands, New Guinea, the South Pacific, China, North Africa, Newfoundland, Iraq, Burma, and, in the case of one lucky corporal, Bermuda.

YANK was an official War Department publication, and all but troops in combat had to pay to get it. The price was five cents, or the local equivalent in a score of foreign currencies. The theory was that no one would read what he received for free. Unlike many other government documents, *YANK* was bought and read because it was believable. In a civilian Army, with all the indignities and frustrations of life under military discipline, it was accepted by the enlisted soldiers and sailors because its editors, writers, photographers, artists, and cartoonists shared the same lot and, if fortunate, bucked their same way up to more sleeve stripes of enlisted rank, but never up to the gold shoulder bars of a second lieutenant.

YANK was designed at the very top of the American war effort by Secretary of War Henry Stimson and President Franklin Delano Roosevelt, as a way to let forcibly inducted civilians compare notes as they endured the separations and miseries of the war effort. One of the few times officers were allowed to contribute was when they spoke up in self-defense in the "Mail Call," replying to some subordinate's blast.

In an era of no television, one mission of the magazine was to entertain. In addition to the pinups, there was always a page or more of cartoons, including Sergeant George Baker's "Sad Sack," the most hapless of the Army's perpetual privates. Sometimes with morbid humor, the cartoonists, many of them non-staffers, laughed away soldier sorrows.

YANK's writers, artists, and cameramen went armed into combat, occasionally manning guns to assist in the battle. The weekly had its share of dead, wounded, and decorated.

YANK is unlikely to be repeated because only in a total mobilization would there be the talent pool from which to draw the staff and the audience of millions for them to address. For those on *YANK* able to continue their careers as writers, editors, and illustrators, and to do it uninhibitedly while wearing soldiers' uniforms, was a sometimes terrifying, sometimes frustrating, often jubilant experience. Perhaps the generation *YANK* portrayed, and within which it served, may not have towered over all others as the greatest. With its patient courage and confidence reflected in *YANK*'s pages, however, it set an example worth following. *YANK* filled its pages with great tales about World War II. As a result, it became one of the war's great stories.

Acknowledgments

This is to thank the late Colonel Forsberg for "designating" me at a *YANK* reunion in 1998 to write the *YANK* story. His attic in a converted carriage house in Greenwich, Connecticut, was a trove of *YANK* internal memoranda.

I wish to express gratitude also to the many staffers and their families for providing generous lengthy essays of their memories and their photos, sketches, and cartoons. Among the many who helped were Annie Davis Weeks in the Bahamas; Janet and Arthur Weithas, who drew on their own research for Art's book *Close to Glory*; Erik Forsberg, who guided me through his father's attic; Ann Forsberg, Franklin's widow; photographers Slim Aarons, Mason Pawlak, and Bill Young; cartoonist Doug Borgstedt; artists Howard Brodie, Bob Greenhalgh, Jack Coggins, David Shaw, and Joe Stefanelli; writers Dave Richardson, Georg Meyers, Ralph Martin, Ozzie St. George, Walter Bernstein, Allan Ecker, James K. Keeney, Andy Rooney, Tom Shehan, Knox Burger, and Robert Bendiner; and, of officers, Major Jack Craemer was especially insightful. Appreciation also goes to Professor Ronnie Day, biographer of the late Mack Morriss, and, in a very large sense, posthumously to Joe McCarthy for suggestions he left for the story of *YANK*.

I thank my son, Martin, for help with the mysteries of the computer and, above all, my journalist wife, Jan, for assistance editorially and in all other ways.

Part One—Staffer

A **1** Racket

"Is it true? Are you on *YANK*? Gee, what a racket!"

It was the spring of 1944 in the Solomon Islands of the Western Pacific, a season of heat and heavy downpours undistinguishable from the other three parts of the year. We were atop "Bloody Hill," hill 260, just east of the Empress Bay beachhead on Bougainville, a few dozen miles north of the earlier fiercely contested island of Guadalcanal. The speaker was a soldier in charge of assembling parties of the "walking wounded."

I conceded that I was from *YANK,* but I chose not to argue about whether my work as a sergeant war correspondent qualified as a sinecure. Ten minutes earlier I had become one of the traffic manager's vertical casualties when a knee mortar shell exploded 15 feet in front of me. A bush of black smoke with a dramatic red heart of flame suddenly had appeared. With my reporter's eye for detail, I stared at the burst so that I could record it, but when I attempted to scribble a reminder there was no dry surface on my pad. The paper was awash in blood pouring from a score of peppery wounds in the face and from a puncture in the chest.

Getting hit caused mixed emotions. The first thought was that I was about to die. Strangely, there was no regret. I was a soldier, and death was an accepted part of that—it had to come sometime. I thought of my young wife and infant son back in New York. What about them? Except for the $10,000 soldier's insurance policy, what further support could I provide them? If dead, nothing. Distressful as the thought was, my survivors would have to find other aid.

But I was still functioning, at least for a while. Perhaps something still could be managed, maybe even time to get off a piece to *YANK*. I took note of how it felt to get struck. There was no pain as scores of tiny fragments pierced the flesh, just a smashing force, as if I had been slammed by a baseball bat. A half hour earlier I had interviewed the Americal Division[*] medics extracting what copy I could from them. I knew where they were crouching. Why not switch roles and no longer ask them what they did, but rather, what, as a patient, they could do for me? They welcomed me back, swabbing away blood and dusting the wounds with sulfanilamide.

A fellow sergeant had been nicked by bits of the same shell. He grumbled angrily. The mathematics upset him.

"The way I figure it," he muttered, "is that the third gets you. This is my second."

The soldier rounding up stricken walkers asked if I wanted coffee. I did. It provided a welcome boost as a group of us started down the steep hill. It was light as we set out, but the sudden equatorial black was already upon us as we scattered on the beachhead below. Then I learned a bit more about woundings. The adrenaline wore off and my legs went numb. I'm going to fall here in the dark never reaching the hospital, I feared. I sent what messages I could to my legs to stay stiff and to keep moving. It worked. I reached the brightly lit field hospital and within moments was spread out on the operating table. A doctor began twanging bits of metal out of my face.

"You're lucky," he said, "It's quiet right now. I have time. If I didn't get these out you'd have blackheads the rest of your life."

One was a quarter of an inch from a jugular vein. Another had just missed an eye.

---

[*] U.S. troops assembled into a unit on New Caledonia in the South Pacific drew their division's name from this: the *Americ*ans of New *Caled*onia.

"I'm going into the chest now," the surgeon went on. "If it hurts, let me know."

I put up with most of the jabs and then grunted, "Ouch, I felt that."

"Not surprising," was the answer, "I'm nearly out the other side."

The medical officer called it a day. He could not find what had caused the larger hole in the chest. Five years later, when I went for an insurance exam, the radiologist was startled to find a pinhead fragment in the heart. He rotated me, took another picture, rotated again, and finally found what he thought must be true. The bit of shrapnel was not in the vital organ. It was just in the heart shadow, but only a quarter of an inch away.

"We'll insure you but with a penalty," he told me. "We'll count you five years older than you are."

Another half decade later the insurers checked again. The fragment had wandered elsewhere. This time the verdict was "we'll put you back to your real age but we won't refund the extra premiums you paid."

I passed the month after the hilltop wounding in two hospitals, first in a dugout on the Bougainville beachhead, then in a more familiar setup above ground back on Guadalcanal. At the first, I watched mangled men across the dirt floor from me waiting to die. At the second, still unsure about the gravity of internal injuries, I used a side room to write the story of the "Second Battle of Bougainville." At a loss ratio of 20 to 1, 7,000 Japanese had been killed, many of them soldiers of the infamous Sixth Division, which had raped Nanking, China, in one of the century's worst atrocities. With illustrations and some textual inserts from my teammate artist Sergeant Bob Greenhalgh, *YANK* gave the story a six-page cover layout, one quarter of the May 19, 1944, issue, just two weeks before the climactic Normandy landings half a world away.

Keeping in mind that we of *YANK* were the chroniclers and not the story, especially when so many others, such as the soldiers in the beachhead dugout, were suffering far worse, I kept my part of the battle account to 7 paragraphs and my comments to 52 words: "(My wounding) served at least two purposes. It demonstrated to my satisfaction that you never know about the one that gets you until the damage is done. And it labored the point that for all the sporting flavor (of our Pacific island) pocket-sized war, it was the real war for keeps."

The reference to the paucity of our island mini-war was suggested by the petty concerns that preceded the shell burst. I had climbed hill 260, not because I saw any story possibilities, but because of what I had taken as a dare. A public relations corporal of the division had asked me whether I wanted to see the close quarter combat on Bloody Hill. I could see no copy that would be sufficient to interest a worldwide soldier audience, including the tens of thousands in England about to invade the Continent. Further, there was a ban on hilltop sightseeing. Kibitzers could only get in the way, risking useless casualties. Still, it smacked of a dare, so, unarmed, the two of us slipped through the front lines and made the 20-story ascent.

YANK and its humorous cartoons were managing to get many a laugh, as well as many a silent tear, out of the miseries of the war, so the hilltop arrival of a *YANK* man touched some funny bones. An excited colonel wanted to show me what he considered an ingenious and hilarious contraption, his own invention, a knee mortar refitted to pitch a five-gallon can of burning gasoline onto a tree stump beneath which Japanese were believed to be crouching. To me, it seemed battle weary nonsense, but I humored the man with the shoulder eagles. We had chatted inside a log-covered command crater as 155-millimeter artillery shells from the beachhead skimmed through the air directly above us with a roar and rattle like that of a freight train.

For the demonstration, four or five of us stepped outside into a clearing half the size of a tennis court, dense jungle concealing all who might be watching from the undergrowth. One blazing can arched toward the stump, then a second. At that point we had a response, answering fire and all of us frivolously wounded.

What got me back to duty several weeks later as a recovering soldier was an order to sweep the hospital ward. Not my job, I objected, my assignment was *YANK* correspondence. Without checking with a doctor, I declared myself cured and walked out, back to duty. With my company headquarters 12,000 miles away on Manhattan's 42nd Street, I was free to do as I pleased, including the option for or against combat. On second thought, I considered that one more week of rear echelon calm might help, so artist Bob Greenhalgh, *YANK* photographer Sergeant Dil Ferriss, and I gave ourselves a recuperation assignment, a ride to Malaita in the eastern Solomons on the Navy's thatch collecting cutter. GI tents were roasting under the tropical sun, while the half-naked Melanesian natives were comfortable under thatch, the primitive, but effective, local version of air conditioning. Gradually, some of our troops were shifting to thatch so more of the straw sheets were in demand.

Malaita, in the eastern Solomons, was only a few years removed from head-hunting and cannibalism, if indeed all the inhabitants actually had abandoned the time-honored custom. Novelist Jack London had written about Malatian savagery and his narrow escape from it. Under sailors' protective rifles, and with chewing tobacco as currency in a moneyless island, we alternated getting photos and copy for a story while trading for war clubs, long slim canoe paddles, bows, arrows, and even baby rattles.

It made a two-page feature for the magazine: "On the Thatch Run, Navy boat tours Solomons to buy grass shacks for GI use."

Being free to wander and cover what we wanted—battles or thatch runs—was certainly uncommon freedom for three Army sergeants, but was *YANK* a racket? Our orders from New York were to send something each week, preferably copy, but at least a letter filling in the home office on what we were doing. I tried to do both, always noting at the top of each weekly report how many weeks had elapsed since I last saw Times Square. *YANK* had a rule that everyone should get out of the 42nd Street headquarters for at least a six-month stint, but when the required half year finished there was a tendency to let those in combat stay where they were. I resolved never to ask to come home, but never at the same time to let New York forget when the calendar pages turned past six months, then nine months, then a year.

In one of the weekly reports, I told the home office of the soldier's remark atop hill 260, my face and chest bloodied, about *YANK* as "a racket." There had been no malice in the comment, rather a sort of reverse congratulations on a fellow serviceman's good luck.

YANK saw it differently and could not have been more pleased. The question was alive whether *YANK* and all the other components of the military's "special service" units were a clever way to evade the battlefield draft. Leonard Lyons in his *New York Post* "Lyons' Den" column had been one of the first to call attention to the anomaly of men in uniform on New York's 42nd Street living through the agonies of World War II while risking no more danger than a possible jaywalking accident.

"Have you heard the news about *YANK*?" was the gist of one Lyons jab. "One of its soldiers just qualified for a Purple Heart! A typewriter slipped off a desk at their 42nd Street headquarters landing on his toe!"

YANK made it a point to fill in Lyons about the exchange atop "Bloody Hill" and, manfully, Lyons ate crow. He told his readers of the ironic battlefield remark not once, but twice.

YANK's staffers, to keep their coverage of the lives of enlisted men and women authentic, were supposed to take potluck with their fellow soldiers. Yet there was always an element of "racket," no matter how hard *YANK* struggled to avoid that image. As I lay on a cot in the makeshift Bougainville medical dugout the day after the wounding, I was issued clean pajamas and taken above ground so that the commanding general could pin on the Purple Heart decoration.

"It showed you were there," he said with approval.

He knew nothing of his corporal's dare, the violation of the no kibitzing order, and the other absurdities that had gone into the exercise.

Meanwhile, beneath our feet in that ditch of horrors, mangled enlisted men lay unnoticed. *YANK*'s wounds were special. That was part of the racket. To get more image mileage out of the Purple Heart, *YANK* printed a letter a few weeks later in the "Mail Call" column. Signed Private G. Saltarelli, and datelined Bougainville, it read:

"Dear YANK,

I have noticed in your magazine that almost every issue has a picture of a famous general. To my idea the greatest general of all has never had his picture in YANK. He is Major General John R. Hodge, former commander of the Americal Division, the first army ground unit ever to take the offensive against the Japs in the Pacific. If you would put his picture in YANK you would not only please me but you would provide interest for every member of the division."

YANK's editors in Manhattan ran the requested photo with a caption they were clearly delighted to include. It read: "Major General Hodge is shown pinning the Purple Heart on Sergeant Barrett McGurn, *YANK* correspondent who was recently wounded by Jap knee mortar fragments on Bloody Hill (hill 260) on Bougainville. It's not a very good picture of the general but our guess is he wasn't bothering much about ceremony."

"Not a very good picture" was a gross understatement. All that could be seen of the commander was the back of his fatigues and the top of his helmet as he twisted around to fasten the award. *YANK*'s own man got far better treatment, a full face shot of my smile and of the large white bandage on my jaw. Pretending not to write about

The author and Sgt. Greenhalgh on Bougainville, just after General Hodge pinned the Purple Heart to McGurn's pajamas. *(Courtesy of YANK)*

itself, *YANK* was pleading its case with soldier readers that it was part of them and no racket.

Yet *YANK*, and the Special Services in general, were a soft touch in the sense that men interrupted by the draft in mid-career were able to carry on their peacetime professions, honing skills, giving them a head start on their life work after the war. Many of *YANK*'s writers, photographers, artists, and cartoonists never lost a step as they went from prewar work, through the years on *YANK*, and then back to higher levels in the same tasks after the victories in Europe and the Pacific. Many others in uniform, by contrast, suffered years of absence from their homes and careers practicing violent skills for which, happily, there was no place in peacetime.

No other corporal and, subsequently, sergeant in history, I am sure, had orders such as Bob, Dil, and I, and some dozens of other *YANK* correspondents, carried into the Pacific. "These enlisted men," the orders said in effect, "can go where they please inside six million square miles of the blue Pacific waters."

Often, it meant that we chose to head for combat. When Japan surrendered in September 1945, I had accumulated so many of the 95 points for discharge—five for each battle, five for a wound, five for a child—that I was a civilian again within a fortnight, while others in fighting units scattered across the world with scant home-bound transportation had to wait abroad well into 1946.

YANK was a racket, yet in a broad sense it was not. It played a needed role in winning World War II. It was the entertainer, confidant, and friend of enlisted soldiers, sailors, Marines, Coast Guardsmen, and merchant mariners, men and women who wondered sometimes whether they had any support. *YANK's* staff understood the anguish of the enlisted soldier because we wore the same uniform, had the same lowly rankings, were excluded from the same officer clubs, and put up with the same frustrations and indignities. *YANK* was the authentic voice of the World War II soldier, and in its "Mail Call" columns offered every serviceman the opportunity to speak across continents in his own voice. It wedded necessary military discipline with the American democratic soul. Without the unreality and distortions of official propaganda, it held up a mirror to the mind of the fighting man, his hopes for survival, his anguish at lost freedoms, his patient courage, and his ability to endure. The GI's mind was complicated, and one that civilians and government officials do well to understand as new generations are summoned to combat.

YANK
Making 2 a GI

We came from many backgrounds.

I was a reporter for the *New York Herald Tribune*, the first on the staff after Pearl Harbor to be accepted as a volunteer war correspondent. I was cleared by the FBI and began marking time until I could cover the first transatlantic landings.

Meanwhile, I continued regular work on two contrasting beats, the Army's effort to enlist enough willing soldiers and the opposing antics of isolationists who wanted no part in any conflict. The anglophile *Herald Tribune* (*HT*) favored the first and deplored the second, but my job was to record each accurately.

The antiwar crowd provided lively copy. In the weeks before the declaration of war, hundreds gathered for angry rallies against taking sides in Europe. Many were members of the German American Bund who had no taste for an assault on former kinsmen. Others were Irish Americans still seething with the memory of historic English repression of Ireland. Another few may have remembered college days in the early 1930s when pacifists at Oxford provided an example to their generation on either side of the Atlantic by pledging never to take up arms, even in defense of their homeland.

As the son of an Irish American father and German American mother, and as the 1935 editor-in-chief of the student weekly at Fordham College, I had been under each of these influences.

A frequent day's work at the *HT* was to take a five-cent subway ride downtown to 90 Church Street, a military headquarters, to check on the progress of the volunteer enlistment campaign. The Army missed its quota every week and, not feeling bellicose myself, I cheerfully reported each failure, ignoring the inevitable consequence, a universal draft in preparation for an unavoidable conflict.

For a sidebar, I rode out to Camp Upton on Long Island where new inductees living in tents on a muddy terrain were coming down with colds and the threat of flu. My story told how poorly the new soldiers were making the transition from steam-heated apartments. Father Coughlin's anti-Semitic and isolationist "Social Justice" magazine reprinted my account, giving it front page play, something I fear that was still unhappily alive in the *Times*'s clippings when Clifton Daniel, at the 1966 demise of the *HT*, told me there was no place for me on his paper!

Father Ignatius Wylie Cox, my ethics teacher at Fordham, no friend of the yesteryear English brutalizers of Catholic Ireland, sent Sylvester Viereck to see me, presumably regarding me as a prospective Irish camel nose under the tent of the London-leaning *Herald Tribune*. Viereck was a main Hitler propagandist. Not happy to meet him, but willing as a curious reporter to speak to anyone, I agreed to quaff a beer in a Brooklyn saloon. Hitler had just swallowed Austria in the Anschluss.

"It's a good thing," Viereck said cheerfully of the Reich expansion. "Let's face it. We Germans are perhaps too harshly Prussian. Austrians are softer. They can have a mollifying effect."

He paused, "Have another beer!"

No, I said. I was happy with my job. Our owner, Mrs. Reid, had no place for Nazis on her payroll. It was the last I saw of Viereck. Even so, I heard one day in the 90 Church Street corridors, "wait until we get him (me) in uniform!"

The threat to let me have it as a private never had a sequel, but within 10 months I was indeed in a medical regiment taking basic training in Abilene, Texas.

At Fordham, our freshman choice was to be on a heavy physical culture program or to join the ROTC—the Reserve Officers Training Corps. With three hours each day spent on the subway commuting from south Brooklyn to the north Bronx, I saw ROTC as the lesser time-consuming evil and soon was doing fairly well on the target range as a sharpshooter.

At the end of sophomore year, we had a further choice: to stay on with ROTC for two more years and be commissioned as second lieutenant in the field artillery, or to quit. Not feeling any more militaristic than the college students of the Vietnam era, I dropped ROTC and bid farewell to an officer's golden shoulder bar.

Within weeks after Pearl Harbor, I was summoned to a meeting with my neighborhood draft board. One member, the father of Ned Cullen, my parochial school classmate, looked on kindly as I announced my *HT* war correspondent assignment and the FBI clearance.

"War correspondent," another board member nodded. "That seems pretty dangerous." The others agreed. It could be grounds for a temporary deferment.

Weeks went by as the country geared up for the conflict. In July, a half year after the assault on Pearl Harbor, Mr. Cullen and his confreres called me back. They needed bodies.

"Why haven't you left?"

"We haven't invaded. I can't go until we do."

My mind and that of the *HT* was focused on Europe, although, about then, Marines were landing in the Pacific on faraway Guadalcanal.

A board member had a thought. Those in essential jobs could be exempted, but how essential is a newspaperman? Not very, to his way of thinking.

"You know," he struggled with his thought. "Let's look at it this way. Say you read a paper on Monday and then don't see another until a week later. What do you miss? Nothing!"

He had a point. The *HT* slogan was "24 hours of world history," and there is only a sliver of that history any newspaper produces on any one day or even in a week. The skeptic of reporters' importance won the argument. I was told, "if you're not on your way by September we must induct you."

I had a suggestion, "Do that, but then let me out when we invade somewhere."

"Oh, no. It doesn't work that way," the neighborhood elders explained. "When you're in, you're in." I was on my way to the Army and, eventually, to *YANK*.

In September, a month or so before the first American troops crossed the Atlantic to invade Morocco, I was summoned for the Army's physical exam. I was determined to pass it. Choosing between an honorable 1A classification and rejection as a 4F, as one unfit, I wanted the first.

"Take one step forward," a line of us was ordered. We obeyed, and with those 12 inches we were instant privates.

Then, in an invaluable preparation for service on *YANK*, I learned what it was like to be an enlisted man, a "dogface," a GI, a disposable item of "government issue."

Other ROTC classmates who had stayed on for the additional two years were officers, one of them making his way to colonel. For those of us who had taken that one step forward we were no longer self-propelled free Americans doing as we pleased and, in such cases as mine, jeering as a newsman at the government's failure to raise a volunteer army. The GI experience was new and sometimes bitter.

What it was like to be an EM, an enlisted man, we learned in a hurry. A first shock was the strip examinations. Any sense of privacy was swept away as we stood naked as cattle for inspection. Then came the "short arm" examination, each man instructed to pull back the foreskin of his penis to show that he had no oozing venereal infection. At the induction center at Fort Dix in western New Jersey, there was the first night in barracks.

"Reveille is at six fifty. We'll call you. Be out in the street in full uniform at attention at seven."

Ten minutes to do that! I thought I had better sleep in my uniform. On rethinking it, I undressed. To my surprise, the next morning I learned how much could be done in 600 seconds. At seven we were all outside, stiffly erect before the sergeant. At Fort Dix, I had my first experience with KP, kitchen police, peeling a parade of potatoes. Thirty-five months later, back at Dix for discharge, there was one difference. Kitchen chores were the work of captured Germans, a mute message to all of us that our war years had accomplished something.

In my 10-man hut in company A of the 37th Medical Regiment in Camp Barkeley in Abilene, Texas, I shared quarters with an Irish American elevator operator from the Hotel New Yorker on Manhattan's west side. Another in our barracks was an illiterate Texan. Endless were the arguments between the big city New Yorker and the man of the once independent Texan nation. Each was convinced of his own state's eminence and, thus, of his own personal significance. At last, the New Yorker had an inspiration, a trump.

"Here's one. In New York, on 34th Street over on the east side, there's a building, the biggest in the world. The Empire State Building. It's 105 stories high." Sharing 34th with his own New Yorker Hotel, the then unrivalled skyscraper was almost his personal possession.

"I don't know about that," was the drawled but quietly confident rejoinder. "What I understand is that there's a bigger one in Dallas!"

Baffled by the Texan's impregnability, and frustrated by life in the Army, the New Yorker went AWOL, absent without leave. He was caught by the military police as a deserter, and was brought back to the regiment a shattered person.

The Texan, too, had trouble. One day our commander, Captain Horwitz, a Midwest dentist, challenged him. "Where's your other uniform? And the extra underwear? And the spare pairs of sox?"

Someone must have tipped off the captain.

"Gone," the Texan said softly. "They got too dirty. I had to throw them away."

Tension between the New Yorkers and the western farm folk existed on other levels. In a two-knockdown encounter with one rangy westerner, I learned that there is literal truth to the scatological expression that one can have "the shit knocked out of him." Thanks to the bulky graduation ring on my left fist, my brother soldier at least got a split lip. Identified as someone with a degree of command presence, and his lip healed, the other was one of the first in the company to don the three stripes of buck sergeant. As one of the few familiar with a typewriter, I became the company clerk, with the two stripes of corporal.

The narrow horizons in so many young minds in cross-section America were brought home as the Army tried to teach us how to soldier and how to succor the wounded. We took a crash course in both, not all with quick success. "Tourniquets are a help but must never be used on the neck," we were counseled. ... Another bit of information was that "we have a score of medicines but perhaps the most important is the last, a placebo, just a sugar pill. If you don't have anything else, use that. The soldier will think you are doing something for him and that may make him better." There was other advice, "If you smell new mown hay and there isn't any, put on a mask. It may be poison gas."

Still, try as they did, our teachers could not get all their messages across. During one dark night exercise, we were told to pretend that we were injecting cocaine, coke, into the injured. An inspector quizzed one private on what he was up to.

"Coke," he said hesitantly. "I'm serving Coca-Cola to the poor soldiers."

In a company of 100, only two of us were college graduates, both from New York. Dave Millman had an A.B. from Fordham's football rival, New York University, NYU. Two months or so into basic training, we were tapped as the company's selections for Officer Candidate School, training as hospital administrators. Dave got his commission, left the medics, and ended the war as the supervisor of athletic programs for the troops occupying Germany.

Before either of us could go, however, our regiment motored east across Texas to conduct maneuvers in Louisiana. As company clerk, my job was to carry messages to widely scattered squads. It meant riding many hours each day, exhausted and sound asleep on the wood floor of a jouncing truck. It was hard bedding but the release it provided was more comforting than any postwar mattress.

Maneuvers provided another taste of the crudity of enlisted life. A bugle awakened us in our pup tents at dawn and a bellowed voice, amplified by a loudspeaker, saluted us as masturbators, "Drop your cocks and grab your sox!"

I had married three months before induction, so my first furlough home to New York was an extension of an interrupted honeymoon. I planned to pass the whole furlough in our kitchen in one of the Vincent Astor railroad flats on East End Avenue on Manhattan's east side, but Ted Cronyn, a fellow *HT* reporter also in service at the time, suggested that I check in at the Army's new Information and Education Service headquarters at 42nd Street and 3rd Avenue. The Camp News Service (CNS) was there.

"You can send in pieces from Camp Barkeley," Ted proposed.

I was already a bylined, unpaid stringer for the local Abilene newspaper, recounting for them the weird experiences of camp life, so I weighed the alternatives. Was it worth two precious hours out of the kitchen? Still uncertain, I decided that it might be. I gave myself two hours to go and return but it was noontime and everyone at CNS was out for lunch. I would not wait, but across the hall I saw a sign, "*YANK.*" Although it was already a half year old, I had never heard of it. To salvage something out of the two hours I walked in.

Captain Hartzell Spence, the editor, was the only person there. I told him I was a newly drafted page-one reporter for the *HT* and a potential member of his staff.

"Can you write fiction?" he asked. It was a skill after his own heart, for his personal claim to fame was that he was the author of *Get Thee Behind Me*, a best-selling book of fiction based on life with his father, a peripatetic minister of the gospel.

Assuming that the correct answer was "yes," I assured the captain that hard news reporting was not my sole qualification. If *YANK* needed fiction, I would oblige.

"Tell you what," said the officer. "What we want is copy. On the side, if you want to shack up with a mulatto, that's all right with us."

I was astonished. I was a happy, monogamous bridegroom and had no intention of establishing a liaison with anyone else of whatever color or race.

Later, I got to know *YANK*'s first editor better. Raised in the devout atmosphere of a rectory, and precipitously decorated with a captain's twin silver shoulder bars, he was trying desperately to be the rough, tough guy and man's man he imagined a soldier to be. His was a misperception and a phoniness that nearly doomed the soldier magazine from its start. Only a year or more later did *YANK* finally find its way as the authentic voice of the bewildered, suffering soldiers. Only with its own staffers deployed far from the 42nd Street headquarters, being shot at, earning combat medals, suffering wounds, and, in four cases, losing their lives, did *YANK* become the true voice and friend of the GI.

Captain Spence made no commitment and, on maneuvers, I had much else to think about. Men over 45 years old had been given the option to return to civilian life. In our company, there were many takers. Others in the company wanted the $10,000 GI life insurance

Author Barrett McGurn shaving in his helmet in front of his pup tent on Louisiana maneuvers, just before "immediate order" to join *YANK*. *(Barrett McGurn collection)*

policy. As time permitted, I had worked on the needed papers. There were many requests still pending in my mind when a telegram arrived from Major General Frederick Osborn, head of the Army's Special Services: "Report immediately to *YANK* in New York."

"Good," I said. "I'll go tomorrow." First, I had to finish up my clerk work. All those who had spoken to me thought that requests were in the works and not just in my head. I had to take care of them.

"No, you go right now," I was told. "Two hours from now there's a truck to Baton Rouge. You must be on it. When a two-star general says 'immediate,' it means immediate!"

There was no time to say good-bye to anyone, not to Dave Millman, not to First Sergeant Jeffcoat, not to Captain Horwitz, not to the transplanted Irish American exile from the elevator shafts of the Hotel New Yorker, not to the laundriless Texan, not to the fist worthy, newly enrolled buck sergeant. I had a secretarial assistant whom I had rarely used. I sought him in a hurry and, while dictating all the unfinished business, assembled my mess kit, knocked down my pup tent, skipped lunch, and prepared to board the truck. Another member of Company A stopped me for a moment, shook hands, and murmured the unmerited but treasured compliment of a lifetime, "I would have wanted to go into combat with you. ..." The Army's secret is to bond, to break down existential individuals, rebuilding them as a brotherhood concerned no longer with self alone. Without ever having been noticed, melding had been underway in Company A of the 37th Medical Regiment.

YANK

Dear God, ③ How Long?

With my writings filling the equivalent of three full issues of *YANK*, I had traveled so much and been in so many battles and theaters of war that as soon as the point system for discharge was announced, after the 1945 Japanese surrender, I qualified for instant release.

YANK had decided during its first months that every staffer would have to spend at least one-half year abroad. Within weeks of its founding, the magazine had drawn jeers that its soldiers were just newsmen, illustrators, and production personnel sitting out a war comfortably in midtown Manhattan while pretending to be part of the fighting forces. For me, the first *YANK* months meant living happily back home with my pregnant bride while going to the office each day to produce such copy as a feature on the women soldiers in the newborn WACs, the Women's Army Corps. While my piece welcomed the ladies to the uniform, I could not exclude a chauvinist hint that perhaps a woman's place was in the home. It was beyond my imagining that, in a nation needing the talents of every citizen, the Army a half century later would have a three-star woman lieutenant general overseeing all military intelligence.

Within six months of the Manhattan idyll, however, I was outbound in late 1943 to the Pacific, the very area my old *HT* had hesitated to cover in mid-1942. Sergeants Mack Morris and Howard Brodie, a writer-photographer team, had completed coverage of the sanguinary battle of Guadalcanal and had been called home. Someone had to take their place in the Solomon Islands, and I was tapped.

Still wearing the two corporal stripes I had received from Captain Horwitz, I took the five-day train ride across the country. Armed with YANK's unique travel orders, I would be free to wander as I pleased through the islands of the South Pacific, but first I had to get to them. As befit a lowly EM, I waited for days at dreary Angel Island in San Francisco Bay until a newly built, unescorted Liberty cargo ship was ready to weigh anchor. With no land in sight for 26 days, we finally made it to New Caledonia, a French holding midway between New Zealand and the Solomons.

The trip provided my first dispatch from abroad. We were torpedoed off the Fiji Islands. It happened at about 10 P.M. one black night as the dozen soldiers on the freighter were crawling into double bunker beds inside two "doghouses" on the deck. I was slipping out of trousers when what felt like a truck slammed into the hull beneath me. It was a dead hit on the engine room. Moments later, there came a second crash. An alarm sounded and we were ordered to the lifeboats to which we had been earlier assigned. There was quiet calm, no one murmured as we tied on life jackets. There was a single exception to the general behavior, a crew member hurried below. I wondered if he would drown there but soon he was back, clutching a cigarette carton.

"Torpedoed once before," he explained. "The thing you miss are cigarettes."

The strange silence continued as the engines stopped and the ship stilled. The first mate ran along the rails peering for holes. We waited for our rowboat ride inside the empty Pacific spaces. The man on watch had reported two white lines lancing toward us through the dark waters but the mate found nothing. The captain came up with new orders. Start the engines again and let's run for it, run, that is, as fast as our lumbering hulk could manage. There were no further shocks. The all-clear was sounded. We hurried down to the mess hall for coffee—bowls of it, not cups. It was much too late for caffeine, but soon thereafter we slept. Two round dents were found in the hull when the cargo was unloaded in Noumea. The submarine may have been too close to arm its torpedoes. There is a minimum arming distance so that the attacker does not sink itself. Or, perhaps we had lucked out and both missiles had been duds.

In Noumea, I picked up weird soldier lore. A spaced-out EM confided about a new intoxicant, gasoline, purified to be sure. To make it potable, he explained, the crusts should be sliced off both ends of a loaf and the fuel should be allowed to drip through the length of the bread. Other soldiers in the South Pacific were already going blind from the wood alcohol in fermented coconut milk. That was bad enough. I did not share the mad tip about gasoline with *YANK*'s thirsty readers.

I was issued a .45 pistol. A sergeant offered to show me how to fire it into a sandbank. Unused to the loaded weapon, I twice rotated it toward my instructor and each time he dived flat to the ground. Getting up the second time, he ended the lesson. For a while, I carried a carbine, but I never fired a weapon. In the Palau Islands, when a bazooka man pointed out a Japanese cave, offering me a shot, I declined the fun. That shot into the cavern could kill someone, man, woman, perhaps a child. The name of the bitter game of war was, indeed, to kill or be killed, but my military specialty did not require me to take life. Absent that, I could not. The metamorphosis away from peaceful civilian ways was taking time.

North of Noumea, on the island of New Caledonia, was a "repl depot," a replacement center filled with worn out soldiers shipped back from combat. I hitchhiked there in the back of an Army truck and listened for hours to frightening stories of night-time hand-to-hand combat with suicidal infiltrators on New Georgia in the north Solomons. It was chilling. I wondered whether these men were shell-shocked and hallucinating. How alarmed would my father be back home when he read such stuff? I decided to write what they said, making no effort to verify the accuracy of their tales, letting *YANK*'s readers judge for themselves. Labeled ambiguously, "Bull Session on New Caledonia," the account filled several pages. Only later did I learn that the "bull" they threw was the simple truth. North of New Caledonia was Guadalcanal. Before boarding a destroyer to head to it and to the still-contested northern Solomons, a fraternal note of counsel arrived from predecessor Morriss, "There are two ways into combat, bit by bit or all at once, both awful."

The dreaded Guadal, by the time I reached it, however, was already a peaceful rear area, so much so that officers were reinstituting the detested spit-and-polish order. The comfortable slovenliness of combat was banned. That made for a chiding story: "Guadalcanal Goes Garrison."

Sergeant Bob Greenhalgh of Winnetka, Illinois, an academy-trained fine artist, joined me as an illustrator, and, during a lull in the fighting, we teamed up in search for features. On Tulagi, an earlier battle area, a grown offspring of cannibals reluctantly told of seeing human flesh roasted in a pit prior to eating.

Trying to draw him out, I asked about favored cuts—the hips perhaps?

"No, I never heard good news about that part," the strapping native answered shamefacedly.

Employed by that time as a handyman at British colonial headquarters, that son of cannibals had accepted the new way of life which London colonizers had imposed. Cannibalism, said the Brits, was murder and would be so punished.

Open-air theaters had been built to show movies. Fallen logs served as seats. Melanesian children clustered in back, staring at the images of urban life. Bob and I did a sketch and text interview with the tots.

McGurn seated, drinking tea, at the British headquarters on Tulagi Island. On the left is the converted son of cannibals, who said, "I never heard good news" about human hams as preferred food. (Sketch courtesy of Bob Greenhalgh)

"What most interests you?"

"The houses," said one child. "Stairs to go up and down on!" His own thatch home not only had no second floor, but the floor itself was made of earth. By contrast, on Florida island across the strait from Guadalcanal, Chief Patrick, barefoot but wearing a battered fedora, showed us with pride his dearest possession and a neighborhood marvel—a wood floor. Bob's sketch of him gave Patrick a moment's global fame in *YANK*.

On the north tip of Guadalcanal, carried there by a coastal schooner, our team of two met one of the local Europeans who, at risk of their lives, were serving as spies on enemy movements. This one, a Dutch priest, Father De Klerk, had been radioing warnings to the troops whenever Japanese raiders flew past him. The priest told us how he had come to this isolated spot. A seminarian in Rome, he had lived there with his mother until ordination and until he was given his missionary assignment at Guadalcanal. Neither he nor his mother knew where the place was, so, on a globe they first hunted for it off Tierra del Fuego at the tip of South America. Finally, they found it, a quarter of a globe away to the west. Arriving at the mission amid Stone Age savages, the cleric found himself as busy practicing unlicensed medicine as preaching the gospel. Along with his missal, he had brought a one-volume listing of basic remedies. There was no other medic there.

Everything about the missioner's life was strange. When he butchered a cow, he would rescue steaks for himself and then watch his parishioners stay up through the night devouring the rest.

At Mass, the priest insisted that ladies, for modesty's sake, slip a blouse over their breasts, but at service's end, as they walked through the vestibule, we watched them strip back to their waists.

Back with the troops, our status as soldiers clashed with our perceptions of what we had to do as war correspondents. It was a phenomenon repeated over and over among all who were on *YANK.* The tropical heat was so fierce that one day the thatch-roofed Navy mess hall at Camp Crocodile burst into flames. All but one of us lined up in a bucket brigade to pour water on the flames. I passed on one pail at a time while making mental notes, but Bob needed both hands as he sketched the colorful scene. With the fire out, we became the target of rage: "You didn't help put the fire out. You can't eat here anymore!"

Calm returned. We did get to eat. But with our unique footloose orders in the face of the Navy chill, we soon considered it prudent to take off elsewhere. Bougainville had been invaded, and a week later a second wave followed. Going with it provided a new war experience. Under cover of an ebony night, the armada moved north through Iron Bottom Bay, so known from the fleet of American warships sunk there in a disastrous engagement months earlier. The nocturnal black was a comfortable shroud until suddenly light pierced the night sky and illuminated us. "What fool turned that on?" a GI asked. It was a Japanese flare exposing us to any enemy lurking ahead. I ducked inside the cabin, out of the way of any flak, and then hurried out on deck again to bob free if the ship went down. Inside or out, nowhere was good. Except for the flare-dropping plane overhead, however, no further enemy appeared and, next morning at Bougainville, we waded peacefully ashore.

The chronic magazine-wide problem of *YANK* men, on how to be soldiers and professional communicators at the same time, came up again immediately. A first job was to dig a foxhole big enough for two. An energetic man of many skills, artist Bob, wielding our assigned ditching tools, soon had a copious space hollowed out. Soldiers with hammocks nearby assured us both of the value of the newly carved ditch and of the vagaries of fate affecting its value.

"There was another guy here who had two foxholes," one said. "Problem was he could never figure out which was safer. One night he hesitated as usual and then jumped into one of them. A bomb landed in the first hole and buried him in the second! We got him out."

It was a good story but were our legs being pulled? I did not file it to *YANK*.

Bob wanted an easel. Among the first support facilities put up on the beachhead was a sawmill. Its output was needed for high priority purposes, but it seemed likely that Bob could scrounge enough if he offered to sketch the mill for *YANK*. He did the illustration and, in due time, it appeared around the world in the various *YANK* editions. Bob got the lumber he needed for his stand. Before the easel could be built, however, a public relations sergeant for the commanding general paid us a call.

Disdainful of my poor two stripes from Camp Barkeley and Bob's three, the general's aide told us that the commander was planning an inspection ride along the front and that there was space in his jeep. We could occupy it if we wished. The way he tendered the invitation, it sounded that the general could care less whether or not two humble EM's came along, so we refused. Hobnobbing with the brass was not our job as we understood it, and Bob had to have the easel. We had to get out copy and sketches. Later we learned, as I should have remembered from that major general's telegram in Louisiana, that a suggestion from a general, whether he has one star or five, is a command. It was months before we heard from the starred officer again. By then, his division was on its way to Luzon and he had been following our wanderings in the pages of *YANK*. He urged us to rejoin him, but by then we were hundreds of miles apart and never made contact again.

Bob and I were especially disappointed in retrospect. With us on that rejected jeep ride would have been Gene Tunney, the boxer who took the heavyweight championship of the world away from Jack Dempsey, the Manassas Mauler, two pugilists who had been our boyhood heroes. Tunney was in uniform as a high-ranking officer assigned to foster morale.

GI life on Bougainville was a mixture of misery and wonder. Eight feet of water falls each year from the sky. Lining up in the jungle rain at lunchtime, we ate hurriedly from our mess kits before the rain turned the mess into lukewarm soup. Skin turned yellow from atabrine as we tried to avoid malaria. Medics painted crotches purple in a struggle against itchy jungle rot and provided no words of comfort—"your real cure is to go to a temperate climate!"

There was a regular nighttime visit from a groaning Japanese bomber, "Washing Machine Charlie." He broke every sleep with what we came to know as our piss call. The whole beachhead urinated into the sand and then jumped into foxholes. As I waited for the whistle of Charlie's bomb, I would experiment with prayers, rushing through the ancient formulas fast to get them done in time and then once again slowly to be sure I had them right. For what good it might do them, I shared my system with *YANK*'s readers.

Slowly the beachhead cemetery grew. There was one report that one soldier, weary of the strain, had shot off a foot, his ticket home.

Values became distorted. Japanese, encamped to the east, had moved within artillery range and lobbed in shells one night as a movie inside a tent reached the best part. It was a dilemma. What to do, rush back to the foxhole as ordered, or wait for the finale and hope it would not take much longer? I chose the latter, and later shared that folly in my weekly travel report mailed back to my first sergeant in Manhattan. He passed the story on to Hollywood press agents and Louella Parsons, then a celebrated motion picture columnist with an audience of millions, published a warm report of how "Sergeant McGurk" had demonstrated the battlefield appeal of what the studios were producing.

In a war there is always someone in a worse predicament who can comfort you by comparison, and another so much better off that it is saddening. The latter was certainly true under our particular jungle trees. The better-off group of soldiers had mounted a steel drum on stilts, filling the container with water, and then enjoying the bliss of a cold shower. We had no access to the contraption but we could strip, walk nude to a newly built highway, hitch a passing ride, and bathe in a river. Fifteen thousand males in an Army division were by then on the beachhead. Not only were there no women, but no man in the division had seen any women in more than a year. When the first nurses arrived, men raced in jeeps to Bougainville's newly built airport to stare tongue-tied at the wondrous creatures of a different gender.

It was the naked river bathing that highlighted problems worse than ours. Beside our jungle hammocks was the encampment of the Nisees, Japanese American soldiers whose job it was to interrogate the few soldiers of Japan who allowed themselves to be captured.

We were American college students, the Nisees told us, just as American as anyone else, although sometimes the girls on campus did treat us as different. Here it is a risk of our lives to bathe in the river.

"A Marine pulled his gun on me," one said. "I shouted 'don't shoot, I'm an American too!'"

It was New Year's Eve 1943, and the gallant Nisees had a treat for us: "The first thing we always do when we get a prisoner is to confiscate his rations. Tonight we are going to have a feast, courtesy of the Emperor. You're invited."

We dined happily, à la Tokyo.

The highways with signs warning "danger, sharp curve," the airfield, the expanding cemetery, the movies under a tent, were developed so quickly, a little America emerged in the jungle so rapidly, that it made a cover story for the paper, "Modern Living on a Beachhead in Bougainville."

There were also many chances to sample different forms of combat. With the Japanese moving in from the east and artillery duels beginning, Bob climbed a tree to sketch the gunfire observers and I flew in a two-seater observer plane for a combined piece on the "Eyes of the Guns." War induces giddiness, so my pilot tried some fun. Turning to look at me, he pressed the Cub into a dive as, over his shoulders, I watched the ocean waters rushing to meet us. Trusting that it was a practical joke, I murmured no warning. The guess was good; just in time the prankster jerked us out of the dive.

The waters were the same ones the future president Jack Kennedy was patrolling as a commander of PT gunboat 109, an assignment that nearly cost him his life when enemy action cut his craft in two. Invited to ride one night on another of the PT patrols, I fell asleep on the narrow front deck, my back braced against the bulkhead. It had been a full day's work ashore on other assignments and I could not stay awake. As I slept, the boat prowled north far beyond the beachhead, hunting for enemy activity.

At 2 A.M., I was jostled awake, "We are under attack."

We lay dead in the water as a Japanese plane circled above us. It curved toward us, squirting bullets 50 yards to one side. In a second swing the stream of pellets was half the distance closer. Did he see us? Was he sure? Was he wondering perhaps whether we were a boat or just a rocky outcropping?

"Next time," said the captain, "we'll fire and run for it."

It was a life-and-death decision. If the pilot had seen us for certain we could at least try to take him with us. If he still had not and we ran, we would abandon our best hope. The PT is fast but it is no match for the speed of a plane. We would leave the same telltale white streak in the water that our *Liberty* ship watchman had spotted when the torpedoes raced toward us off Fiji. The plane would follow that trail, catching us the way a clammer uses spit marks in the low-tide sand to find his buried prey. We waited in silence. The plane veered off, giving up the attack. He had not been sure after all.

It made no story for *YANK*. Soldiers in Europe, the Middle East, Burma, and New Guinea were living through scarier threats to their lives and there was nothing else I knew about the PT mission in the Solomons to pad out the story.

What did make an article, however, was a combined New Zealand and American assault on an atoll north of Bougainville. General Douglas MacArthur, in his memoirs, explained the importance of little Green Island. Seizing it and building an airfield in less than a week's time in effect ended the battle for the Solomons and Bougainville. Japanese still entrenched in jungle pockets could be left behind to wither and die as the Allied advance swept north. Like our fellow EM, Bob and I were not privy to that high strategy. All we knew was that we would be part of another invasion. We slept for a few hours on a transport and were awakened at 5 A.M. by the siren call to "general quarters," battle stations. The sky was already filled with the puffy black pockmarks of exploding shells as American and Japanese planes twisted in dog fights and warship guns found targets. Dozens of Japanese enemy planes spun into the ocean, leaving

Greenhalgh's cover describing our eerie walk with New Zealanders through the freshly captured jungle of Green Island. *(Courtesy of YANK/Bob Greenhalgh)*

long black smoke trails standing in the air above them. Bob's sketch made it to *YANK*'s pages.

By 9 A.M., the order to land came. It was tense as our landing craft poked through a narrow passageway into the broad lake at the heart of the circular atoll. We were an easy target for small arms fire, but no sound came from the green jungle wall. Inside the lagoon, our cup-shaped landing craft circled and then pushed through shallow waters toward the opaque emerald curtain of the shore. The front side of our small boat rattled down as a ramp, and we dropped into water above our knees. What awaited us behind that vegetal wall ahead? Happily, there was no resistance. By 11 A.M., there was a memorable treat. Anzaks, as we called the troops of New Zealand and Australia, had a bonfire blazing and on it a huge kettle. From it came a lifetime's heartiest, most bracing drink of tea.

No sign yet of the enemy. We joined a patrol going single file through the dense forest. We passed a hastily abandoned camp. An officer's hammock was still suspended from vines. Open-sided thatch-roofed lean-tos for enemy EM looked too damp for American comfort. A half mile into the eerie stroll, with fears of ambush dogging each step, we reached a bombed-out mission, the enemy headquarters. It yielded a trove of booty and souvenirs, antiaircraft guns, radio sets, gas masks, an officer's leather case with a fountain pen, wristwatch, and clothing, and supplies of rice and canned goods. Unlike the fight to the finish that the Japanese had put up on other islands, this garrison must have been ordered to flee. Bob's sketch of the jungle walk made a full page cover for *YANK* on May 4, 1944, one month before the Normandy landings.

With little more than a platoon of *YANK* staffers covering all the area from California to India, the 42nd Street headquarters ordered me back across 1,000 miles to cover a major strategic conference in Hawaii. The imperious Douglas MacArthur was driving north toward the Philippines and the equally proud Admiral Chester Nimitz was island-hopping from the east. They were converging. Only one could command. Who would it be?

Technical Sergeant Art Weithas, one of *YANK*'s few five-stripers, was on hand to do sketches. The inventor of the *YANK* format, and the magazine's art director ever since the original dummy issue, he had come west in line with the weekly's rotation policy. Also on hand was one of the few Navy EM on *YANK*, Chief Petty Officer Mason Pawlak, a photographer, who now succeeded artist Greenhalgh as my teammate and illustrator. Bob went on to other island combats.

The three-man conference on whether to drive past or through the Philippines on route to Japan was, of course, top secret, and we know now that there was no discussion of the atom bomb which would end the war a year later. We know now that MacArthur got the lion's share of the Western Pacific controls, but at the time I neither knew nor cared just how the divisions were made. That was hush-hush major strategy, no concern of any man in the mud. Still, from *YANK*'s worm's-eye view, that was not my story anyway. Marines hidden in the Waikiki bushes granted an interview on how they managed to look so sharp. Each man stood a four-hour hitch as a presidential guard, using the rest of his waking hours to spit-shine his boots, soak cartridge belts in brine, to sharpen their appearance, and rub linseed oil into rifle stocks for a high mirror glow.

The cheerful crippled president received the press for a few words on the broad beachside Waikiki lawn but only, unwitnessed by us, after he had been carried to a bench from which to greet us. A three-striper by that time, I was delighted to know that the commander-in-chief understood the markings on my sleeve. "Hello, sergeant," he grinned.

YANK's idea was to give its staffers as much as possible of the life of the civilian war correspondents, so Pawlak and I shared a rented Honolulu apartment rather than going into Army barracks.

Free of military limitations, it was a good staging area from which to cover feature stories, such as a photo and text account of what famous baseball players, now EM, were doing as soldiers in Hawaii.

Staff Sgt. Joe DiMaggio, after his blow-up with McGurn. (Courtesy of YANK/Mason Pawlak)

Private Joe Gordon, of the New York Yankees, posed good-naturedly mopping a latrine, and *YANK* published my verdict: "Despite his scrubbing, the porcelain would not have passed inspection."

The reaction was different from Gordon's teammate, four-striper Staff Sergeant Joe DiMaggio, world known for the longest streak of base hits. At 2 P.M. one day, we found him comfortably asleep in his barracks. Far from the dangers of combat, he was doing his bit in the national defense by swinging a bat in ballgames organized to entertain fellow GIs. Startled at the uncommon sight of a soldier enjoying a mid-afternoon doze, and anxious to wrap up the story, I made the mistake of nudging the celebrity awake. Joe not only refused to pose, he raged, "I'm going to tell the general on you!"

37

The champion baseball player, no doubt, was as good as his word for, days later, with Pawlak accompanying me, I was shipped back to combat and *YANK* was offered material for a two-page layout explaining why DiMaggio's ball playing and that of other EM drafted from the White Sox, the Reds, and the Browns were a contribution to the cause. It made a good story but, unlike other *YANK* copy, it carried no byline. Headlined "Report on Joe DiMaggio," it said that Joe "probably realizes better than anyone else that he is on the spot and that to most GIs in combat zones his job smacks of special privilege."

Not like that at all, the text suggested. It was not Joe's idea that he play ball but rather that of a team called the Seventh Air Force Flyers, which "looks on all major leagues as its farm." The work was no cinch either, six games a week in two Hawaiian leagues, one of servicemen and one including some civilians. For the layout, Joe not only consented this time to pose, he did so five times. There was no latrine shot, to be sure, but there was one of him marching in parade, another of him with needle and thread sewing on his stripes, and one at the players' training table with him dining from a plate rather than a mess kit. The caption explained that DiMaggio was "wolfing" down "the same chow as in any enlisted mess." In what some military PR man and ghostwriter may have taken for soldier talk, the layout, as a topper, cited a nameless GI who "growled": "A lot of fellows think it's a picnic out there. Jockstrap soldiers! I'd like to hear somebody call DiMaggio a jockstrap soldier. I'd slug him in the eye with a bat."

Soldiering is mainly shared suffering and combat. My mistake was to think of it as solely that. Like *YANK*, itself inside the same "Special Services," there were support units making a contribution not always involving the risk of death. Joe was providing R and R, rest and recreation, a recharging of the spirits for the others who were temporarily out of the fighting. His contribution was real, even if others doomed to lie under white crosses and stars of David did not always see it that way. Joe stayed playing ball in Hawaii for two years.

Meanwhile, Mason Pawlak and I sailed back west with a dozen civilian correspondents aboard the heavy cruiser *Portland*, the *Sweet Pea* as its sailors called it.

On the *Pea* we had a quick reminder that our job not only was to see things through EM eyes, but that we were EM. The civilian reporters were berthed in gentlemanly comfort in Officer Country. Mason, as a chief petty officer photographer, the rough equivalent of an Army five striper, fared not unbearably in cozy chief quarters while I weighed the choice of a sailor's bunk or sleep on a desktop. Faithful to our peculiar *YANK* logic that we were a somewhat different kind of EM, I chose the latter. The desk as a bed was no worse than the truck floors on maneuvers in Louisiana, and I slumbered well. The one negative was going topside in the morning, climbing ladders up claustrophobic shafts as narrow as pipes. With scores of others trying to funnel their way up the same slot at the same time in an abandon ship exercise, I wondered how horrifying that would be.

Food was no problem. It is Navy lore that the chiefs, supervising rations for all onboard, officers and EM alike, dine best of all, with the captain himself a possible exception. Mason got us both into the chiefs' mess and mealtimes were a pleasure.

The days steaming toward the setting sun provided an opportunity to learn about the lives of sailors aboard a cruiser. In two years the *Pea* had crossed the Equator 18 times, converting pollywogs who had never before left the Northern Hemisphere into shellbacks, veterans of travel both under the North Star and the Southern Cross.

"Last year," offered Yeoman Third Class Paul Carpenter of Heathland, Alabama, "every place we went wasn't ours or wasn't ours until we got there."

Pea had plenty of "gedunk," sweets on sale at a soda fountain. For 10 cents, there were ice-cream sundaes with chocolate syrup but, for some reason, the day I checked, no chopped nuts. Silex coffee-pots were everywhere, even in the big gun turrets. There were at least 15 jukeboxes aboard, ready to blast out canned music.

For 15 cents, Seaman First Class Luther Winkler of St. Louis sewed on new "ratings," the sleeve indicators of rank. For a dime more, Winkler pressed pants. During combat, he put aside his thimbles and needles to help with damage control.

There was only one "plank owner" aboard, Chief Otis Rutland of Vallejo, California, the sole member of the crew to trace back to the 1932 commissioning. Youngest, already a veteran of two years of service, was 17-year-old Seaman Third Class Donald J. Martin of Jeffersonville, Indiana, who had lied about his age to enlist at 15. Eldest was Chief Bo'son Mate John "Pop" Hughes, who looked forward to saluting his nurse ensign daughter at her commissioning two years hence.

Squiring us west, the *Pea* already had behind her a distinguished record of combat achievement since the Pearl Harbor disaster had launched the war. Of the first eight major battles, she had missed only one. With the midnight sun of the Aleutians throwing off sleep patterns for some of her crew, there were complaints of a different nature only a fortnight later in the Tropics as the thermometer below decks in the fire room registered 178 degrees F. For one 99-day stretch, there was never sight of land, and for 105 days, the *Pea* never entered a port.

At the battle of Midway, the enemy air attack went on for a full day, the enemy losing 125 planes. An even greater strain was the battle for Guadalcanal in November of 1942, when the *Pea* was at "general quarters," full battle status, for 57 hours. Her guns shared credit with the *San Francisco* for sending a Japanese battleship to the bottom in three minutes, but a Jap "can," a destroyer, returned the favor, shooting off the *Pea*'s tail with a torpedo.

Chief James R. Henington of Douglas, Arizona, was topside

when it happened. "It felt like going up an elevator," he said. "Then down again. We were trying to get the hell out of there but we were just going in a circle." Down below, Chief Howard Selden of Tacoma, Washington, had a different take, "It sounded like the main guns, a nine-gun salvo, only three times as loud. It knocked the men in the main engine room to their knees."

The *Pea* limped to Australia for repairs that time, but was soon again back in the fighting.

How did the sailors bear up under strains such as sustained air raids? "After a while," said Chief Henington, "they don't have any significance. Everybody gets scared. You can't say they don't. But finally they get used to it. You're afraid and yet it's an everyday thing to you. It's hard to explain."

Seaman First Class Donald Lawson of Louisville, Kentucky, had his own answer, "I wouldn't trade my experiences for all the money in the world, but I wouldn't want to do it again either."

Ashore back in Guadalcanal, by then a major staging area, Pawlak and I boarded a Marine troopship heading for a third invasion, this one in the Palau islands north of New Guinea and east of the Philippines. As the *Pea* and others bombarded Peleliu, it seemed as if the island would sink, but instead the first wave faced heavy fire, which, intermittently, was still going on a month later. Which wave to take in? To write about it, I reasoned, the sixth was as good as any and, hopefully, less lethal. It was one of *YANK*'s blessings. Not under orders like other soldiers, we could choose our own poison.

The night before landfall, I looked into the Marine quarters. Silent 19-year-old young men, many of them destined never to be 20, were bent over rifles, giving the weapons a final cleaning. A chief petty officer eased my anxiety. He knew how to get into the ice-cream store and we enjoyed some.

Ashore the first night, I shared a foxhole with one of the Marine Corps combat reporters. He alarmed me with his foghorn snores.

Occasional shells crashed dozens of yards away, and after one of them, there was a scream of agony followed quickly by silence. I feared that my companion's honking sleep sounds might bring us enemy infiltrators and, indeed, sometime after midnight, I was awakened by gunfire 20 feet away as an alert Marine killed a creeping enemy.

Pawlak and I, anxious not to come under Marine command, eschewed any official tenting, and instead built ourselves a wigwam out of raincoats. Every night it rained in, and the next day we would requisition another poncho to cover the new hole. The Marine supply sergeant was endlessly obliging and unquestioning, so we ended with a shelter of 20 raincoats. We still rose wet every morning.

It was battlefield lore that the Army and the Marines took about as many casualties, the cautious Army collecting them slowly over time and the Marines in a frontal assault swallowing them all at once. The differing Army–Marine mentality, plus the tension of combat, got our two-man *YANK* team a death threat one day.

"I could have you executed," a ranking Marine officer raged at us.

It was not about the misappropriated raincoats, and he explained.

"You were out on the ships and said that the Marines are throwing lives away!"

Then I remembered. Toward the end of the fourth day of the invasion, we were told that there would be a press conference aboard the U.S.S. *Fremont.* Usually, we skipped such meetings. Our job, as we saw it, was to chronicle the life of the "grunts," leaving strategy to the brass. But this meant a momentary escape from the horrors of the shore, so we grabbed the chance. As Pawlak recalled a half century later, we were thrilled to be out for a few hours sitting "among famous journalists, high ranking Navy and Army officers, while drinking freshly prepared coffee and being served sandwiches and even cold beer that dripped water on the green cloth of officers' tables instead of the miserable droplets of our perspiration saturating Marine combat greens unchanged in four days."

What neither of us knew was that a controversy had been brewing between foreign correspondents who had never gone ashore and Major General William Rupertus, commander of the Marine First Division. Rupertus had told the press that although the Palau island of Peleliu would be a tough and dirty job, it would be over much faster than the recent costly fight at Tarawa. Offshore, the Army's 81st Division had the 321st and 322nd Infantry waiting to help, but they were released instead to invade another Palau island, Angaur. The Marines were doing Peleliu by themselves, though they had to admit they were taking many losses. "The commanding general," an Australian correspondent said, "is a glory seeker. Frontal assaults kill too many men."

Eyes turned to Pawlak and me. Unlike so many others at the heated meeting, we had just come from the beach and, what was more, we were GIs, the only ones present to speak for EM. Our job was to listen to all sides, learn what we could, and then express our-selves in our reports to *YANK*, not to a press conference. We did our best to stay silent but, involuntarily, we nodded at some of the protests. Pawlak had photographed casualties "literally stacked up on the beach," as he later put it. When the Aussie blurted out his accusations, Pawlak thought of a deal he had made a day or so before with another of the Australian civilians, Damien Parrer, a newsreel photographer. Parrer had been with us on the *Sweet Pea*. To stay out of one another's way, he and Pawlak took opposite flanks in an assault. Parrer died over there under machine-gun fire.

I had already seen enough to know the high cost of head-on attacks, although it was nearly a month later on Peliliu that I saw the consequence of one order. Marines had been sent up the steep side of Bloody Nose at the left side of the beachhead. Most of the troops bogged down partway, and weeks went by before anyone could climb all the way unchallenged. With the way finally clear, I ascended the ridge and, 50 yards beyond the crest, found the remains of two Marines who had managed, perhaps at the start, to make it all the way. Alone in the shallow valley behind the Nose, they had been picked off by fire from a cave just ahead. Their uniforms were as sparkly clean as if they had just come from a laundry. Constant downpours and blistering equatorial heat had washed and dried them. Inside were skeletons.

Back ashore, the morning after the press conference, Pawlak and I were taken into custody and marched to a Marine command post. In proper military fashion, we stood stiffly at attention as the officer finished other business. Then he turned on us in a fury, accusing us of telling the world press that the Marines were squandering lives. We were guilty, in his view, of an assault on Marine morale. "I could have you executed," he said.

In retrospect, the remark was preposterous. An already difficult Marine public relations problem would have been magnified, to say the least, if the leathernecks assassinated a couple of *YANK* men. Still, the two of us took it seriously. Life had become cheap on the battlefield, and the exhausted officer was unstrung. Fifty years later, Pawlak says that he is still haunted by the episode. "In war," he notes, "anything can happen, whether things be real or imagined. It still goes through my mind. 'Demoralizing the Marine Corps ... Twenty Years Sentence.' ... Or, 'Shot For an Act of Treason.'"

It was no time for an EM to argue with a superior, but we quietly denied that in conversation, or in our *YANK* reporting, we were bad-mouthing the Marines. We were dismissed, "We'll see. You can't leave the island anyway." So the irate officer thought, but we said good-bye to our raincoats and hitched a quick ride to the Army on Angaur where Pawlak soon had a closer call.

A Japanese cannon was on rails inside a cave. It would come out, fire, and withdraw. Already it had immobilized a tank and a second armored vehicle had moved up to replace it. The first tank had to be blown out of the way. An infantry column was moving up single file

Pawlak's cover photo of Angaur. In the foreground is the Japanese corpse that a soldier stepped over, saying, "The only souvenir I want to take back is my ass." In the background is the tank where Pawlak was knocked unconscious. (*Courtesy of* YANK / *Mason Pawlak*)

to help. We fitted into it. One by one we stepped over a Japanese corpse. On his chest were handsome binoculars. Booby trapped swords, ceremonial bellybands, and other battlefield goodies had already blown off many American arms and legs, so it was no surprise when the soldier in front of me glanced down, shook his head, looked up, and muttered, "The only souvenir I want to take back is my ass." In an era of language niceties, even for GI publications, I cleaned that up for *YANK* as "my rear end."

Off to a side was an enemy bunker, a thick layer of coconut logs above a narrow ground-level firing slot. No one objected as a bulldozer sealed the slot, burying alive any soldiers inside.

Pawlak got an idea for a good series, the step-by-step process for knocking the stricken tank out of the way: first, the picture of the tank, next, a close-up as the demolition charge was set, finally, a quick run back to safety behind the second tank and a shot of the exploding wreck. The first two in the series came off well but, just as the charge was set, a sniper hidden in a tree opened fire. Pawlak was pinned down next to the exploding charge. He was knocked unconscious. I decided that we had had enough for one day and brought him to the medics. He survived but, for 10 years after the war, had recurring fevers and still has pains across the left side, which faced the explosion. There is partial loss of function in his left eye.

With Pawlak's photos, the story of the two Palau island battles, and of the massive naval bombardment that had preceded them, filled two *YANK* pages under a stoic title ignoring the personal horror and focusing on the big picture, "Power Play at Palau."

YANK

Join 4 MacArthur

YANK's six-month-abroad rule had been satisfied in my case twice over when *YANK* in Hawaii asked *YANK Down Under*, in General MacArthur's command, whether help was needed as the invasion of the Philippines came near. The Western Pacific *YANK* may have had other reporters in mind, but the reply was "Yes, send McGurn."

In the Palau islands, "Join MacArthur" was the terse instruction Pawlak and I received from Hawaii. Join him, but where was he? His Southwest Pacific domain stretched many hundreds of miles from urbane Australia through prehistoric New Guinea. Taking literally that we had to find the general himself, the two of us hitched a ride to Angaur's freshly built airfield and found a pilot who was heading south.

"Do you know where MacArthur is?" we asked. "Are you going there?"

He had no idea of the general's whereabouts. Presumably that was secret. "I'm doing a milk run east along New Guinea ending back in Guadalcanal," he volunteered.

"Fine," we told him. "Can we hop off at each stop, see if MacArthur's there, and get back on if he isn't?" That was agreed and, at the first landing in north-central New Guinea, no one had any idea where their commander was.

Next stop in New Guinea was Hollandia. MacArthur had captured it by following his master plan to hit the enemy where he wasn't.

Strong enemy points were leapfrogged, leaving the foe to wilt in his rear, while he skipped farther to other weakly held spots. None of the targets meant much by themselves. Strength had to be preserved for the enemy homeland, the real objective.

"Is MacArthur here?" we asked. "No," was again the reply. "Not down here at the field, but see that bungalow up on that ridge? He's there."

Our blind EM approach to the solution of a strategic problem had worked!

The first thing was to run back to the plane to grab our mess kits, bath towels, steel helmets, a carbine rifle, cameras, a typewriter, and the rest of our gear before the plane left for Guadal.

On the ridge we were greeted with dismaying news, "You missed it! You should have been here yesterday! The ocean was covered with ships as far as you could see!" That was it! We had to be on that! "Well you missed it, but there will be a second wave six days from now."

Not good enough. We had to go now.

There was a compromise. An armed but unescorted black widow flying boat would take off in a day or two to carry the general's mail. We could ride in that. We did. Groaning west along the jungle coast, we landed at night on freshly captured Morotai Island, some miles east of the habitat of the legendary Wild Man of Borneo. Living conditions were austere in the first week of occupancy, but the island boasted a large cache of sake, the potent Japanese rice wine. Mindful of the dreaded booby traps, I drank water and left the sake-quaffing for another day.

Memories of past battles, and the thought of more to come, caught up with me that night on Morotai. "Dear God," I wondered, "how long will this go on?" For me, it was a momentary low point, but not as bad as some others had suffered. I remembered a camp, well hidden from impressionable serviceman eyes, into which I had stumbled in the New Hebrides south of Guadalcanal. A naked man was in a cage made of the perforated long metal slabs used as surfacing on newly built airstrips. The cage was in a small compound being used as a trans-shipment point for the evacuation of deranged soldiers.

"He was one of our staff," a person in charge said. "It all came on slowly. The first sign was when he stopped opening his mail." Perhaps exposure to others with psychiatric disturbances had overturned his own emotional balance. "It gets cold here after dark," his comrades said. "We give him a new blanket every evening. He tears off a one-inch strip and then another and another until nothing is left. Then he goes naked through the rest of the night."

I chose not to recount it to *YANK*. One reason was that all of us on *YANK* had to go through censorship, although I found the censors marvelously benign, concerned mainly with protection of the secrecy of upcoming engagements. Mainly, I had no context. I had no information on how widespread were similar psychological breakdowns and what the Army was doing to comb the victims quietly out of the ranks. *YANK*, however, at just about that time, on March 31, 1944, did come up with a three-page spread on war fatigue. Written by Sergeant Morriss, whom I had succeeded in the South Pacific, the article included no statistics on how many soldiers had at least temporarily lost their minds, but it did make the point that the GI who understood why he had to be in service was least likely to succumb.

On D-day plus one, the second day of MacArthur's long-promised return to the Philippines, our load of mail took off from Morotai and, a mile or two offshore, we flew up the coast of the big Philippine island of Mindanao. Tall columns of smoke rose inland, presumably targets struck in the softening operation before the landings farther north.

In Leyte Gulf, we saw what we had missed at Hollandia, a steel-covered ocean with one part of it the *Sweet Pea*.

Landing in the water, we took a small craft to one of the larger

McGurn and Pawlak in MacArthur's mail plane flying from New Guinea to the Philippines. (Courtesy of YANK/Mason Pawlak)

vessels, where I was told I had to go immediately ashore. Boats were for sailors, Mother Earth was for a soldier. On the beach, I learned how much can be done with a helmetful of water. You can wash your face, take a shave, do a complete body bath, and wash out sox and other clothes. You can try to wash a white bath towel with what liquid is left, but, unhappily, it will always come out brown. A few *YANK* staffers from *Down Under*, including Sergeant Ozzie St. George, the best-selling author of *Care of the Postmaster*, had set up magazine headquarters in a log-topped dugout, but some of the Southwest Pacific crew already had a better idea. General MacArthur had his eye on a comfortable residence in the town of Tacloban, and his colonels were at work commandeering other properties for the high command. One of the more imaginative *YANK* staffers had a moderately felonious idea. The Japanese police had just fled from a two-story frame building two blocks from MacArthur. A Filipino flag was hung from one of its second floor windows. No *YANK* sign was posted. It looked as if a patriotic local family was in residence. The fraud worked. No staff major general ever found it, and *YANK* had comfortable quarters for the duration of the Leyte battle.

The departed police either had been handy with shovels or had Filipino labor to help them, for the building came equipped with an ample foxhole out front suitable for a half dozen *YANK* men. One night, three weeks after the landings, the enemy evidently decided, if possible, to eliminate our neighbor, General MacArthur. There were 20 raids between 10 P.M. and just past midnight. In an earlier attack, our house had become a sieve of shrapnel holes as a bomb killed a civilian correspondent a house or two behind us, but this night none came that close. Even so, it was exhausting. We jumped out of the mosquito-netted beds bequeathed to us by the vanished gendarmes, ran down the stairs, and jumped into the hole, all of it quick enough so that some other *YANK* body would serve as the lid on top. At plunge 20, I made a decision. Life was sweet but there was a limit to how much it was worth. I had to sleep. I would run no more. Happily, the 20th raid proved to be last.

General MacArthur set up an enlisted-man chow line on his front lawn, and the food there was good enough, but we found an appealing way to supplement it. Soldiers guarding the food dump were happy to load a jeep with foodstuffs for us in the wan hope that their names might show up in our pages. What they gave us was all but inedible boxed rations, but a family across the street from us was happy, in exchange for half of our supply, to use native spices to make our half something to swallow. We were free also to use the family's sanitary facilities, a short perforated wooden bridge over a stream through their front lawn. Thus did the Army's official magazine manage to work and survive.

Wandering along the beachhead in search of a story, I stumbled on one of the turning points of the Pacific war, while failing to understand it.

"It was amazing here yesterday," a soldier said. "There was a 14-year-old kid who crashed his plane 100 yards off the shore. He said they taught him how to take off but not how to land."

He was one of the first kamikazes, suicide bombers, who killed hundreds of sailors and sank many warships in the months just ahead. I had only one man's account of what had happened, and no inkling as to what it meant. *YANK* missed a scoop.

In New York, my "Log of the Sweet P" was still sitting in the pile of unused copy. *YANK* asked me for a new lead, a newsy update. I found that the *Pea* had been one of the bombarders of Leyte and sent that off as a substitute first paragraph. The "Log" got page one billing and filled two pages. But again, from an EM's worm's-eye view of the conflict, the story of life aboard the heavy cruiser missed a much bigger update. The *Pea* was in the midst of one of history's greatest and most decisive sea battles. The Japanese Navy had converged on the Leyte landings in an effort to destroy all ashore and the American Navy sank it, bringing close the defeat of Japan. Like many another Leyte EM, I learned only after the war how close we had come to being a morsel in the mouth of the Nipponese dragon.

Sixty weeks after leaving our 42nd Street headquarters, I was called home to take charge of *YANK's* national bureau in the Washington capital. A reason may have been my Purple Heart. Critics had taken note that, by no means, all of the millions in uniform were much in harm's way. Congress was grumbling. One remark was that "if all the officers and enlisted men were pulled out of the bars of Washington we could put another 15,000-man division into the field!"

As a starter, an order was put into effect that any sergeant, corporal, or private in Washington for more than 48 hours had to move into Army barracks. For *YANK*, the edict was unacceptable. It went

YANK's Philippine bureau on D-day plus three, on Leyte Island. From the left, Maj. Hal Hawley, Sgt. Dick Hanley, McGurn typing a story, best-selling author Sgt. Ozzie St. George, and Sgt. Chuck Rathe of YANK Down Under. (Courtesy of YANK)

across the grain of the magazine's two-year campaign for a life and working conditions for its staffers as close as possible to those of their civilian opposite numbers. Like civilian publications, *YANK* had its bureau in the National Press Building. Maybe a Purple Heart on the uniform of their Washington man might deflect some lightning as *YANK* proceeded with its usual scrounging.

New York staffers on a scanty per diem cash allowance were living in civilian quarters. That gave them free movement as they covered the home front news. The Washington man needed the same. How was I to be kept out of one of the Washington forts? The answer was in the small print. Forty-eight hours! There was the solution. Do a day's work in the capital, get a round-trip Pentagon rail pass to Baltimore, check into a hotel for the night at government expense, and then head back to the capital for another 36 hours. Keep that up until *YANK* could get an exception. That required only a week but, in the meantime, the scheme almost blew up, putting my assignment in question.

Dining at the home of a man just under Cabinet level, I was asked how my new job was going. Indiscreetly, I mentioned the rule-dodging Baltimore excursions. My host left the table abruptly. "Where's he going?" I asked the son of the family. "To call Drew Pearson," he told me.

This was a good story for the columnist muckraker, how one branch of the Army was thwarting another. Jobs as an enlisted man living in the relative comfort of the capital were hard to come by, and I was about to lose one of them. "Tell your father this was privileged conversation," I urged the son.

The host returned. The scandalmonger wasn't home. The man of the house agreed to desist, and *YANK* had survived another of the perils of publishing a world magazine with none but enlisted men and a woman or two as staffers.

YANK

Back 5 Home

YANK sought to be all things to fellow soldiers, not just sharing the excitement and horrors of combat, but all other interests as well.

In most stories about the human condition, boy finds girl, but in the Pacific, some Army units and ships at sea went more than a year without seeing a woman. When the first nurse landed on Bougainville, soldiers raced in jeeps to look at her. Back home, she may have been a plain Jane, but on the beachhead tarmac, she was so thrilling a sight men were too timid to speak. A picture is no substitute, but it may be better than nothing, so, for soldiers, YANK had its pinups, often studio shots of Hollywood actresses. They were torn out, fastened to barracks walls, and copied in paint on the noses of bombers.

Even the most hardened infantryman sometimes needed a confidant, a receptive ear or a counselor. YANK's "Mail Call" column and its answer man met that need.

By the start of 1945, some soldiers were completing their required service and faced the longed for, but intimidating, return to civilian life. Others, in Europe, with the collapse of Italy and the reversal of Nazi fortunes, faced a transfer to the Pacific and a new kind of warfare.

For all these, YANK had answers, and as the newly assigned chief of the magazine's national bureau in Washington, it was my new job to provide many of them.

The first order of business, after giving up the round-trips to Baltimore, was to find a flat where I could be reunited with my wife, Mary, and infant son, Bill. Getting Pentagon permission to stay out of enlisted barracks was one thing, quite another was to find an apartment. The sleepy prewar town of Washington had been converted overnight into the throbbing heart of a global war machine. Where to stay? Apartment seekers were snatching up the first editions of the newspapers, racing through the classified ads, and then speeding to newly posted lodgings before others could beat them to it. One man offered to squeeze the three of us into his cramped quarters if my wife would pitch in as his cleaning woman. Others said I was moments too late. Finally, one officer who was quitting town on military orders said he was "awarding" his flat to a serviceman married to a servicewoman, a doubly worthy couple, but that he had an idea for me. The man upstairs had just received his own departure orders and had no time to tell anyone. The neighborhood was undesirable, along the railroad tracks, but I grabbed the apartment.

Housing solved and the family back together, a "racket" indeed that throngs of my fellow GIs would have coveted, I got my first homeland assignment. The United Nations was about to be founded in San Francisco. It would put an end to wars! Fresh back from combat in a world still at war, I felt that the idea was well intended but fatuous. No more wars—how could soldiers still under fire believe it? Rather than take it seriously, I skipped the solemn proceedings and instead did a feature piece on what civilian negotiators from abroad thought of American women. Well aware of the hunger of overseas troops to be back with their wives or girlfriends, and how eager they would be to trade places with the out-of-country diplomats on California's peaceful streets, I asked delegates provocatively what bones they had to pick with America's ladies. More than a few, was their innocent answer.

Farid Zeineddine from Syria busied himself with nitpicking.

"Too much gum chewing," he said. "And too many wear glasses," he added. "Perhaps the glasses," he surmised, "came from staring up the sides of such big buildings."

Next to join in was Creighton Burns, a foreign correspondent from Australia. He was appalled by the full-throated profanity of one woman conductor he had encountered on a cable car. She had railed at a passenger. Most inappropriate, he felt, though he had to admit that those aboard were in general "a noisy, pushy, discourteous lot!"

"You Americans" was the title for the two-page spread. *YANK*'s gifted cartoonist, Sergeant Ralph Stein, provided the illustrations, happy-go-lucky bespectacled secretaries peering aloft while tugging strands of gum.

The piece represented my reactions back home in the first days of 1945, with the fate of the war still uncertain but, in retrospect, I know that my treatment of the fledgling UN was too cynical. The new world organization was no joke. It did not abolish wars, but it may well have headed off a few.

Next came stories from across the country, spelling out step by step how newly released soldiers were using generous government loans to resume civilian life. Though the Asian war still raged, some GIs had completed their share of the fighting, and *YANK* knew that millions still in uniform hoped one day to follow in the same footsteps.

Private Roy Rufus Hayes, an ulcer sufferer discharged from the 504th Military Police Battalion, described in Linden, Texas, how he became the first veteran to receive a GI loan to buy a farm. He had to go both to government agencies and to a bank to put together the $2,000 needed to purchase 65 acres of woodland—ten acres for cotton, ten for corn, four for peas, and two for sorghum. Coupled with plans for a few cows and a mare to breed colts, it struck the banker and the federal officials as a deal worth even as much as $600 more than the $2,000 the seller was requesting. The government chipped in half the $2,000 at 4 percent with full repayment due only in 1965.

Anyone interested in getting his own farm was able to use Roy's story as a road map on how to go about it.

A few miles from Washington, in Falls Church, Virginia, Jack Charley Breeden went through a similar exhaustive checklist as he told how he picked up the first GI loan for starting a business. It had been no cinch.

Though I had never met him, Charlie had been aboard the destroyer *Philip* in our small flotilla at the capture of Green Island. He was a torpedo man, mate first class, and was another ulcer honorable dischargee. Before service, he had been a plumber's helper and, before that, had studied for three years as a "little general" in the military program of the Washington and Lee University preparatory school. Teaming up with a 4F (unqualified for service) friend who had run a meat market, he decided to go into the business of carting carcasses from slaughterhouses to butchers. For that, he needed a refrigerator truck. If he could get a bank loan for half the cost, then the GI bill would provide up to $2,000 for the other 50 percent.

"A lot of guys think they're gonna walk down the street and say 'that's a good little gas station, I think I'll get the government to buy it for me,'" Breeden said. "It's not like that. It's no joke, I'll tell you that." It was bad news on the face of it for any soldier looking toward a self-employed future. As Breeden listed everything he had been required to do to convince a bank it would be repaid, and to assure government bureaucrats that his venture had a reasonable chance, his story of how he answered each question demonstrated how it could be done.

Soon in 1945, GI business loans were outnumbering those for farms by a margin of 3 to 1, and similar grants for housing were outdistancing farm purchases 41 to 1. That called for a visit to the two veterans who got the first GI home loan. Forty-one-year-old Private First Class Herbert J. Pugh had been the military policeman on duty at the gate of the camp at Cherry Point, North Carolina, where

Corporal Florence Strong outranked him but chose nonetheless to accept his hand in marriage. Pfc. Herbert got an over-age discharge, and section 858 of the Army code (pregnancy) freed Florence. The Pughs found a $7,200, one and one-half story suburban house in Herbert's home area of Richmond, Virginia, and the government agreed to guarantee one-half of an $8,000 bank loan since each qualified for a $2,000 GI loan. However, conditions were no less onerous than those Hayes had faced when he bought his farm and Breeden had to overcome to get his truck. The couple cleared one of the hurdles when Herbert assured the banker that he was a local, not someone likely to move away, but then the bank and the Veterans Administration came up with a barrage of questions.

Would Herbert's weekly earnings as a salesman of automobile parts be able to meet the monthly mortgage payments? On a 4 percent 20-year loan, that came to $59.50. The answer was yes; with commissions Herbert was averaging $60 to $65 a week.

How close was their church, the grammar school they would likely need in six years, and the high school in 14?

Did the house fit into the neighborhood, neither too good for it nor too poor?

With part of the loan meant for repairs, could the couple handle the $555 the Veterans Administration considered necessary on floors, walls, plaster, and woodwork? The Pughs said they were sure they could, and they were astounded later to discover how close the VA appraiser had come to the mark. Their actual bill for the repairs was $568, just $13 over.

Putting down only $500 of their own, the former private and the ex-corporal got the house, and soldier readers could tell from their story just what they could expect as home seekers should they outlast the war. It was possible to get a loan, but it was by no means easy. Months after the Pughs' success, a study of 9,421 other veterans who tried for the home loan showed that only 934 survived the paperwork.

Only they went through with a request. They applied for a total of $4,211,367. Only a third, 305 of them, got anything, and even they had to share a mere $187,305.

The series on returning veterans and on the GI loans filled nine pages. The Veterans Administration reissued the loan stories in booklet form as an easy way to explain the piles of red tape, and to encourage the uncertain to believe that there were happy endings at least for a few.

While the loan stories spoke of a possible jubilant civilian future, the truth for millions still in uniform was the ongoing war. Defeat of the Germans still left the Japanese. A new assignment was to tell the victorious troops in the cities of Europe what to expect in the equatorial jungles and on the coral reaches of the Pacific, and then later, in an assault on the homeland, of a foe who customarily chose death rather than surrender. It was May of 1945. With what I intended as straight talk, soldier to soldier, I mapped out three pieces. First, how good a soldier is the slight-statured Japanese? Very good indeed was the answer. Next, if he doesn't kill you will the tropical diseases do that job? They could. Finally, the jackpot, how long, dear God, must this go on?

For an answer to the third, I bussed across the Potomac to the sprawling Pentagon. One colonel, who should have been in the know, remarked that "you can't tell, it could take another year or two."

Thinking of the kamikaze suicide bombers, the Japanese troops scattered across the Asian mainland, and the idea of a block-by-block advance across Japan, I thought the man with the shoulder eagles was bonkers. I looked for another.

"No way of knowing, it could take 10 years," said Colonel Number Two.

More like it, I thought, but I gave it one more try.

"Impossible," the third colonel agreed with the other two. "You know in history we have had a Hundred Years' War. This may be another."

Thinking back to Peleliu and Angaur, where defenders chose to be buried alive or to die in the fire of flamethrowers, I thought that the third guess might be the better of the three.

New York published the three articles, but before the mats could reach the most distant edition, there was no war. Like Vice President Harry Truman himself, my three colonels were not among the few with a "need to know" about the atom bomb.

In the spring of 1945, the White House had become an important *YANK* beat. At V-E Day, the victory in Europe, I asked for an exclusive presidential letter of congratulations from the fading Franklin Roosevelt. It would be a scoop and it would serve as a bookend for a similar FDR message in the magazine's first issue, but I could not imagine anything the occupant of 1600 Pennsylvania Avenue could say usefully to comfort those who had seen comrades slain and still dreaded a similar fate. The President had played the most crucial role as commander-in-chief but, in personal terms, no bullets had cut the air on Pennsylvania Avenue. Mine was, I confess, a short-sighted GI view of the burdens of leadership as well as the high risk of being the chieftain in a lost conflict, something spelled out so in the fate of FDR's foes, Mussolini dangled dead upside down in a Milan gas station, and Hitler, a suicide.

The White House agreed to my request for a presidential statement, and startled me, "You write it!" I wanted the scoop but, mindful of the anguish of so many of the GIs heading for new battlefields in the Pacific, I could think of nothing comforting the White House could say. *YANK* missed the scoop.

Weeks later, the gallant President Roosevelt was dead. *YANK* opened 13 of its 24 pages to the loss of the war leader and the arrival of his successor. The burial was in FDR's home in Hyde Park, New York, so I was asked to get permission for a staffer to ride the funeral train. The White House refused. Space was too limited, just enough for news agencies, a few major newspapers, and little more. FDR's son, James, passing through the West Wing pressroom, saw my gloom and asked why. *YANK* was off the train! The President's son walked back inside and moments later returned. "You're in," he said.

But it was not quite that easy. The hard-pressed White House press office still had a card to play. I was a *Herald Tribune* man on temporary war duty. "If we put *YANK* on, we have to bump Bert Andrews from your *Herald Tribune*," a press officer said. It was not just the *HT* but it was Andrews as well, an old friend. In 1939, at age 25, when I had first gone abroad as a *Tribune* foreign correspondent, the sage Andrews had given me some wry fatherly advice, "Be sure you don't carry your wallet in your back pocket."

But *YANK* claimed my loyalty now. "Bump Bert," I told the White House officer. *YANK* rode the train.

Part Two—Voice of the Soldier

YANK

Thumbs Up ⑥ from On High

The revolutionary idea of a major military morale instrument under the control of sergeants, corporals, and privates, with no say-so from the generals, was born, only partially, in the mind of a World War I enlisted man, but it flourished because the most glittery of the big brass gave it support.

Egbert White was the erstwhile enlisted man with the idea. He had served on the brief-lived *Stars and Stripes* newspaper of the American expeditionary force in France in 1917–1918. When the 1941 Japanese attack on Pearl Harbor plunged a hesitant America into the then two-year-old conflict, he had his vision. He was vice president of Batten, Barton, Durstin, and Osborne (BBDO), a premier New York Madison Avenue advertising firm. People were volunteering for service as officers in the rapidly expanding forces, and White saw a way to help and for him to fit in, this time as one of those in command. Less than a month after the declaration of war, on January 6, 1942, through a mutual friend, he sent a suggestion to Brigadier General Frederick H. Osborn, head of the Army's morale branch, urging that what was needed was "a well-organized general newspaper with special editions for various areas and expeditionary forces." White said he was willing to work on such a publication, but not as one of the "dollar a year" men, well-to-do persons of talent serving unpaid. White's assets were not such that he could afford that but, with an officer's pay, he was ready to "make a heavy financial sacrifice in order to make a contribution."

Osborn was interested, so two and a half weeks later, on January 24, the ex-GI sent him a detailed "plan for a service of information for the military forces."

To help flesh out his thoughts, White formed two informal committees; the first was made up of former EM staffers of the Paris *Stars and Stripes*, the second of magazine and newspaper publishers.

With a new war stretching across Europe, north Africa, Asia, and the Pacific, it was clear that the soldier editors and writers would have to put out a more or less timeless magazine, readable around the world, not an up-to-the-minute daily newspaper. *Stripes*'s EM staffers in the first conflict had worked under the control of the local military commanders as agents of their purposes, but it was evident that a single global periodical would have to be out from under local oversight. Still left open was the position to oversee the new periodical's EM.

White's committee of former *Stripes* enlisted men included many who enjoyed communications careers between the two wars: among them Harold Ross, founder and editor of the sophisticated *The New Yorker* magazine, Grantland Rice, a celebrated sportswriter and creator of the Football Hall of Fame, Mark Watson of the *Baltimore Sun*, Alexander Woolcott, a radio celebrity, and Franklin P. Adams, whose column, "FPA," was a main feature of the *New York Herald Tribune*.

One of BBDO's business contacts in the magazine world was the popular weekly *Saturday Evening Post*, so, for his second committee, White tapped R. M. Fuoss of that publication's leadership. Others he chose included Alfred C. Strasser of the weekly *Liberty*, and another veteran of *Stripes*, Adolph Shelby Ochs of the *Chattanooga Times*, a member of the *New York Times* ownership family.

General Osborn was impressed with a still more–detailed proposal White was able to give him by January 24, but he asked for further specifics before taking the matter up with General George Marshall, the Army's chief of staff and number one military officer. With the help of his committees, White, on February 25, produced 36 pages of recommendations.

First, they had a name. Call it *YANK*. The name "is short, punchy, easy to say, memorable, and colorful."

Next, there were ideas on content, some of them aimed at the Army's needs but so far removed from the reality of the life of a World War II soldier that the mature *YANK* later ignored them. One was that *YANK* make heavy use of stories on enemy atrocities so as to "make every soldier want to 'get one of the bastards' before lunch." *YANK* later did run some pieces of that sort, but they did not become part of a regular editorial diet.

Another proposal was to "give the soldier the impression that he is the healthiest, best fed, best equipped fighting man in the world," and that "he is going to be a chump unless he gives every ounce that's in him." To its own subsequent embarrassment, *YANK*'s first issue on June 6 did indeed follow this guidance, pointing out that $50 a month base pay for an American private was second only to Australia's $62.50, and more generous, sometimes by light years, than Canada's $30, Germany's $21.60, Britain's $12.20, Russia's $4, Italy's $1.51, Japan's 30 cents, and China's 20 cents. Neither the White team nor *YANK*'s first soldier editors focused on the absurdity of thinking that a man in uniform could draw comfort from the thought that he was abandoning a sometimes lucrative civilian career and endangering his life for a base pay of $1.62 a day, even if that was $1.42 more than his Chinese opposite was pocketing each month.

"The first job of this paper would be to arouse the fighting forces to desire to fight like hell (and) therefore the lead story should always be a fighting story," the White group urged, adding that "we must not be too holy to engage in occasional leg art."

Both points were featured subsequently in *YANK*, combat

stories because that was the essence of soldiering, and "leg art," or pinups, not run just occasionally, but instead for a full page, in every issue, 4 percent of the tabloid's contents.

General Osborn was sold on the project, so Egbert White, the erstwhile enlisted man, was leapfrogged over the ranks of lieutenant, captain, and major, to be sworn in as a lieutenant colonel, charged with the task of producing a dummy issue for inspection by Secretary of War Henry Stimson, and, as it turned out, by Mrs. Stimson.

Osborn submitted the layout to Secretary of War Stimson, who liked it, but when the war secretary took it home Mrs. Stimson protested. That pinup! The secretary's wife had stern ideas about what should be on the minds of the boys in uniform, and half-clad ladies were not on her list. No problem, the White team hastened to reassure the secretary of war. The offending lovely was stripped from the dummy and winsome actress Deanna Durbin, swathed from top to toe, became the substitute pinup.

White needed still another team to provide the nuts and bolts of a new periodical. Happily, the solution came from a conversation in the men's room at the *Street and Smith* magazine offices. As they washed up, two salesmen were chatting about what they had heard about former *Striper*s trying to start a new publication. Hearing them as they gossiped was Franklin S. Forsberg, a man in his mid-30s, publisher of the company's *Mademoiselle* magazine. A fluffy periodical for girls in their teens, *Mademoiselle* was not the sort of periodical whose readers would be out "getting bastards before lunch," but Forsberg did know how to produce and sell a magazine. He was sure he would be drafted, and he thought he could do better for the Army as a publisher rather than toting a gun.

The eavesdropping introduced Forsberg to a magazine he would ultimately command. He saw no incongruity in the locale where he had picked up his knowledge. An instinctive friend of *YANK*'s future EM staffers, he laughed in one postwar interview, "We used to say about the Pentagon that the only place they knew what they were doing was in the restrooms!" Forsberg sought out White, was commissioned a major, and became *YANK*'s business manager.

Mademoiselle had a skillful master of layout, Art Weithas, who, after the war, had a large hand in designing the Elizabeth Arden cosmetic luxury look. Inside *Mademoiselle*'s art department, taking a lead from the February 24 recommendations and working straight through for 36 hours, Weithas and the others came up with a 24-page dummy layout for *YANK*. To a great extent, it was precisely the way *YANK* would appear for the next three and a half years.

General Marshall was engulfed in mobilization problems, but he was brought into the discussion. There were many questions, not just the matter of who would oversee *YANK*'s enlisted men. Why, for instance, did the Army need a magazine when the newsstands already offered so many well-done publications, such as the *Post, Colliers*, and *Liberty*, each of them selling for a mere 10 cents while *Newsweek* was available for just a nickel more? Might there be an entrepreneurial backlash against the Army if *YANK*'s $50-a-month privates offered cut-rate competition to the costlier staffs of traditional periodicals?

There were still more concerns. Would Congress object to financing *YANK* as a part of an already inflationary military budget? How could bulky shipments of *YANK* get to troops scattered across the globe in timely fashion? Would soldiers have to buy *YANK*, or would it be free?

Both Stimson and Marshall were among those asking questions.

By April 22, four months into the discussion, Osborn, now the ball carrier, had some answers. As for whether there was a need, "there is a general complaint from the camps about the lack of reading matter." Civilian magazines were not providing the information and entertainment wanted by the newly inducted soldiers. Sales in the military stores, the post exchanges, proved the point. With 2 million men already under arm, the usually popular *Colliers*, for one example, was selling only 35,000 copies a week, one for every 50 servicemen.

Even if the civilian publications were to sell, would that help? Osborn doubted it. Such periodicals sometimes go in for "constructive criticism," which is not very "inspiring" for the troops, he said. Also, he added, they frequently run articles and publish pictures "of a kind not useful, perhaps instead actually hurtful, to an Army spirit." He had an example. Photos in *Life* magazine currently on the newsstands showed gruesome Bataan gangrene cases. As the officer in charge of troop morale, Osborn saw the shots as "pictures I would not like to see generally circulated in the Army."

Left open was the question of whether *YANK*'s future soldier editors would consider a frank discussion of war's horrors the only way to win EM confidence and, in that way, to sustain morale. Osborn wanted readers to pay for *YANK*. "Americans," he said, in answer to that question, "do not like free publications. They are suspicious of a free paper. A free Army paper might come to be regarded as merely a government handout." To make the GIs pay, admittedly, was the hard way to do it, putting on the staff "the burden of publishing a paper the soldier likes well enough to buy," but there were counter advantages, he wrote. For another, it would head off "the possibility of criticism from Congress and the public for this use of government funds."

On the question of how *YANK* could get to the far corners of the world, Osborn gave assurances that "lightweight stock for airmail distribution" might be an answer, unless the future staff could come up with a better one. Later, puzzling over it, they did.

War Secretary Stimson gave his consent, but with one proviso, no distribution of *YANK* inside the United States, at least at the outset. That would help with some of the political problems. It was the green light. A four-page typewritten document gave the go ahead. Signed in Marshall's name by Colonel John R. Deane, secretary to the War Department general staff, it concluded with the magic words:

Thumbs Up from On High

"Approved by Order of Secretary of War, April 24, 1942."

It was less than a third of a year since White had come up with his concept. The first hurdles were passed, but many missteps would occur before *YANK* emerged as the most successful publication in the history of the Army, and as the pioneer of global publishing.

YANK

A Bastard **7** Before Lunch

YANK began with two readerships to serve: fellow enlisted men and the big brass who had midwived its birth. At the outset, unsure yet what to do about the first, the latter dominated.

Although *YANK* was to be the voice of the enlisted man, the first two aboard were officers, soon to be followed by a third. Colonel White, who had thought up the publication, served as overall commander, with Major Forsberg at his side serving as business manager. Editorial contents are the sole purpose of a publication, so the two tapped United Press writer and part-time novelist Hartzell Spence as the captain to handle that.

From the outset, it was accepted that the official voice of the enlisted man needed some officers. They had to be the bridge to the rest of the Army, to handle administration and to oversee personnel and finances. How the EM would run the editorial contents would be decided later on.

The next questions were where to set up headquarters and how to pay bills. Anxious not to invite protests from Congress that appropriations were being frittered away, the Army told Forsberg he was on his own. The Army would provide soldiers for the staff, but that would be it. On his own tab, business manager Forsberg had to float a $25,000 loan from the Army post exchanges for use as start-up money. Eventually, he not only paid it back, but had a profit of

$1,250,000 as well. At war's end, he handed that into the Treasury.

What helped with Forsberg's difficulty was Osborn's early insistence that readers would have to pay something if *YANK* was to be credible. He ruled out running paid ads, the usual source of a periodical's revenue, so the copy charge was crucial to survival. The decision on how much to ask was still up in the air when Weithas designed the *YANK* cover. In his view, the price had to be part of the layout. *The New Yorker's* Harold Ross had been helpful in so many other ways that the designer went to him. He had a concern. Should he just say five cents? Ross could think of a dozen worse difficulties in launching the magazine so he laughed, "You do have a problem!"

With a little more thought, five cents was chosen as the magic number. It was half that of the major weeklies. With *YANK* soon phenomenally successful, it paid the bills.

YANK's offices moved in with the Army's morale division in the Bartholomew Building at 205 East 42nd Street in New York's midtown. It was a good publishing location, in the heart of the city's printing industry. Still, it was a far cry from a military camp. Savoring the irony, *YANK* staffers soon referred to the location as their "Fort Bartholomew."

YANK's writers and illustrators took up housing in nearby hotels. A per diem pay of $6 comfortably covered lodging and restaurant costs.

As the designer, Weithas trusted that he, too, would be commissioned. Now in his eighties, he remembers that Colonel White "guffawed" at the suggestion. Weithas was sworn in as a private, but with the magazine organized as an Army company its "table of organization" was wide open at the start. There were provisions for several first sergeants so, within a few weeks, vaulting over corporal and buck sergeant ratings, Weithas had a first sergeant's five stripes, and the $114 a month base pay that went with them. Months later, one officer, who was a stickler for soldier protocol, learned that First

Sergeant Weithas had never gone through basic training, the school for marching, saluting, and the other niceties of GI life. Thus, the magazine's designer was pulled off his job for several days for a hasty reintroduction to the Army at Governors Island, a New York Harbor fort a short ferry ride from the south Manhattan Battery.

Fleshing out the editorial original staff of 19 were eight privates; Bill Richardson, the nattily attired former Sunday editor of the *San Francisco Chronicle*; one single striper private first class; one two-stripe technician fifth grade; four double-stripe corporals, including photographer Pete Paris who was to die in the Normandy landings; four three-stripe buck sergeants; and Weithas as first sergeant.

A rough dummy of *YANK* on April 4 had military superiors, rather than common soldiers, in mind. Far from the real world of the 1942 soldier, the mock-up tracked on the White committee's recommendation that *YANK* should inspire the troops to "get one of the bastards before lunch." The dummy displayed the slogan "by the men for the men in the Army," which was carried in one form or another throughout the periodical's life, but the contents were lurid. With fictitious news events as the subjects, the cover used inch-high type to call attention to an inside "picture story" of "the rape of Bali," as well as to a page five account of how a "*YANK* rubs out 81 Japs." The Bali spread covered two pages, headed: "Exit Bali, Once the Island of Tropical Splendor." One picture, captioned "Japs [*sic*] idea of fun," showed dozens of half-naked women and children sprawled dead on a staircase. "Rape of mother and child" was the identification on another. Graphic shots of wounds from mustard gas and dum-dum bullets padded out the display. Most of the remaining pages were blank.

By June 6, a fleshed-out free issue was ready for limited distribution. "Sample copy, for overseas circulation only" was stamped on the cover. Much of what *YANK* became as the true image of the WWII GI was in the issue, but the initial weaknesses of professional civilian-hearted journalists posing as fighting men were evident.

The issue carried Sergeant George Baker's first depiction of the Sad Sack, a pathetic, anti-heroic, sparse-haired, large-nosed, big-bottomed, slumped-over, skinny character who, through the course of *YANK*'s life, became the epitome of the suffering soldier, a walking joke who made hundreds of thousands of dejected draftees feel better by the sight of someone whose condition was worse. For many, the Sack was *YANK*'s greatest reverse contribution to fostering good morale among the troops.

Marion Martin, a blond Broadway showgirl, was the full-page pinup. Only her head and shoulders were shown, but if any reader missed her sultry, mascara-eyed look, the caption called it to his attention.

The cover story, as recommended by the White team, was a war account, an exultant report of the Royal Air Force obliteration of much of the cathedral city of Cologne. "Auf wiedersehen (farewell), Cologne," jeered a subhead.

The features "Mail Call" and "The Poets Cornered" made the first of many appearances. With no readers yet, the staff made up letters, answering them with what the newly inducted editor took to be GI language. "Pvt. Marvin Wilson" wanted to know whether it was true that "the worse a gas smells the less harmful it is," phosgene resembling hay and tear gas apple blossoms. Wrong, was the reply. "What about mustard gas or do you like the smell of garlic anyway? And ethylichlorasine which has a biting odor and blisters the bejaysus out of you."

Corporal Arden L. Melott, of the headquarters company of the 772nd Military Police Battalion, tried his hand at verse:

The Bugler
The bugler wakes us up each day;
I wish to heck that he would play
Over the hills and far away.

In addition to Baker, four other soldier cartoonists contributed, inaugurating a series that later ranked with the pinups among *YANK*'s most popular features. Masked in humor, they portrayed the absurdities and hard truths of military life. One cartoon by Private Bill Mauldin, who did not stay with *YANK* and joined the *Stars and Stripes* newspaper, where his description of the anguished lives of the disheveled Willie and Joe made the Pulitzer Prize–winning Mauldin the war's premier soldier cartoonist, depicted a throng of admiring civilians hovering over two bandaged GIs. "I haven't the heart to tell 'em," one whispers to the other, "that we got this way at the roller rink."

The White committees had urged that each issue include an editorial supporting the war effort. With Captain Spence as executive editor, and Corporal Harry Brown of the Engineers as acting managing editor, the newly inducted soldiers, still by instinct civilians and professional journalists, tried their hands at being tough-guy spokesmen for their fellow dogfaces. With mocking references to shavetail second lieutenants and to COs (commanding officers), the effort was awkward at best. Entitled "*YANK* Reporting for Duty" it read:

> The Army is made up of a bunch of guys who regard themselves, after a while, as men who know something about this business of fighting. Soldiers, all of us, look down on a rookie or a newly minted officer with professional disdain that makes the worldly-wise attitude of a Broadway lounge lizard tame by comparison. The mark of a veteran should be a lifted eyebrow rather than a service stripe. ... Today, *YANK* recruit, reports for duty. ... There's going to be hell to pay before this thing's through. ... *YANK* is your paper. ... It's not interested in shavetails or the CO—it's interested in the finest men who ever rolled a pack and hit the field. Move over, soldier. You've got company.

One would have gone far to meet soldiers "lifting eyebrows" at the most foppish of "Broadway lounge lizards," but at least *YANK*'s newly assembled staffers were giving their task a try.

YANK

Pratfalls 8

As the authentic voice of the World War II enlisted man, *YANK* had a bad attack of laryngitis as it uttered its first words.

Its newly inducted staffers tried to think and write like soldiers, but efforts were strained, and some of them brought down the wrath of the very brass who had agreed that the GI staff should be free to publish as they pleased.

The magazine went on sale for the first time on June 17, 1942, with a send-off so spectacular that even the most level-headed would have found it hard not to be giddy. President Roosevelt himself contributed the text for the whole of page two, written on stationary labeled "The White House, Washington." Through *YANK*, FDR addressed all servicemen abroad:

> To you fighting men of our armed forces overseas, your Commander-in-chief sends greetings in this, the first issue of your own newspaper.

The President said he accepted fully that *YANK* should be the soldier's voice, "a publication which cannot be understood" by the enemy. He added:

> It is inconceivable to them that a soldier should be allowed to express his own thoughts, his ideas and his opinions. It is inconceivable to them that

79

soldiers—or any citizens, for that matter—should have any thoughts other than those dictated by their leaders. … I look forward to reading *YANK*— every issue of it—from cover to cover.

The veterans of *Stars and Stripes,* who plugged for *YANK*'s creation, entertained the newly assembled staffers at a testimonial dinner at Manhattan's Lotos Club four days before the first issue, and arranged for a nationwide radio cast of part of the program. On the broadcast, songwriter Irving Berlin introduced soldiers who were performing in his show, "This Is the Army," and General Osborn, the Army's morale chief, read FDR's *YANK* tribute.

A printed program included salutes from civic and military chieftains of several nations. Secretary of War Stimson hailed "this publication edited for and by soldiers." General John J. Pershing, the World War I commander, wrote in that he was too feeble to attend the event, but he predicted that, just as "no other agency did more (than *Stars and Stripes*) to sustain the morale of our troops in 1917–1918," *YANK* would accomplish the same in the new struggle. General Marshall, who had had his own share in fashioning *YANK,* said that it was a "source of great pleasure to me" and would "provide great comfort" to the troops that *YANK* was on its way. From abroad, Charlotte, the Grand Duchess of Luxembourg, while expressing thanks to General Pershing for liberating her country in the First World War, predicted that the postwar files of *YANK* would be "a kind of Golden Book of American gallantry and heroism." From the United Nations High Command, Field Marshal Sir John Dill commented that a publication "by and for the enlisted man" with staffers who are "both good soldiers and good newspaper men" could do much "to keep up the spirits" of the troops.

It was heady stuff for a staff of EM, almost half of whom were privates, but the troubles of the real world came down upon them as soon as the first issue went to press. Congress had just approved a

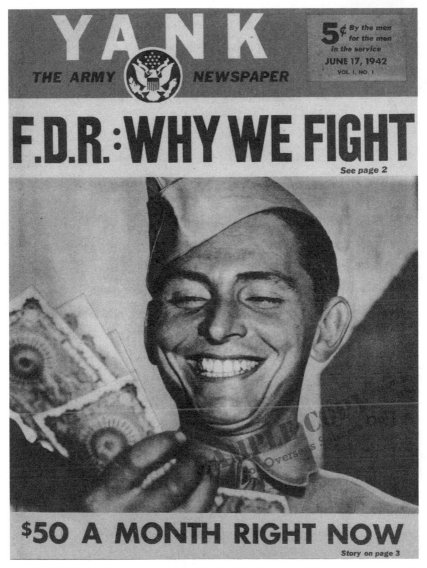

YANK's infamous first cover, which had to be destroyed because it suggested that American soldiers were mercenaries. *(Courtesy of YANK)*

raise for those under arms, privates jumping from a pay of $21 a month to $50. That struck the art editor as a good full page cover shot, so Sergeant Johnnie Bushemi, a staff photographer and former field artillery man, asked Private Homer Oliphant, a peacetime song writer and a new addition to the *YANK* ranks, to pose holding aloft a fistful of dollars. Oliphant did, grinning broadly as if $1.66 a day struck him as a king's ransom.

Thus far, no problem, but the art department failed to check with acting managing editor Bill Richardson who had just received FDR's presidential letter. In it, the President not only greeted *YANK*, but he also used the opportunity to urge all under arms to realize that they were fighting "for a life of peace and decency under God" and for "your homes, your families, your free schools, your free churches (and for) the thousand and one simple, homely little virtues." Happy with the exclusive presidential message, and unaware of the nature of the cover picture, Richardson's assistants slapped a line above it, "FDR: Why We Fight, See page 2."

Meanwhile, just as contented, the art department pasted its own banner caption under the currency-waving Oliphant, "$50 a Month Right Now, Story on page 3."

Read at a glance, the first page of the first issue of the magazine "by the men for the men in the service" seemed to be saying, along with no less an authority than that of the commander-in-chief, that the volunteers and draftees in uniform were happy mercenaries, little different from the hated Hessians who had fought the patriots in the Revolution of 1776. Fifty thousand copies were run off before anyone noticed the appalling juxtaposition. All but a handful of the 50,000 were scrapped. With the Lotos Club gala reception only hours away, and the American Army's war effort still only in the planning stage, the harassed art department hastily replaced the Bushemi shot with an action picture of a field artillery crew. A tough-guy caption replaced the mention of the pay raise. It read, "Spoiling for Action. U.S. gun battery in Australia ready for foe. 'Let 'em come,' they say, 'we'll murder 'em.'"

Apart from whether artillerymen would talk like that, the caption also stretched another aspect of the truth, Weithas admitted later.

The soldiers in the picture were Australians.

The cheers for the birth of the soldier's magazine were not over. Major Forsberg's old publication, *Mademoiselle*, wanted to do something. After all, *YANK* saw the light of day in their art department. The fete took place in the elegant skyscraper rooftop Rainbow Room, one of New York's most sophisticated venues. It was too much for the Marines's *Leatherneck*.

"Woo woo, *YANK*," the rival combat magazine jeered. "What are your thoughts on hairdos, perfumes and fashions?" the Marines asked *YANK*'s soldiers.

The White committee had recommended fighting editorials for each issue, and acting managing editor Bill Richardson's staff provided them until, a few months later, they caused a crisis. The editorial in the first issue was in a self-conscious staccato style as tartly abrupt as the name of the magazine itself. There were 20 paragraphs, 17 of them a single sentence, and the other three with only one sentence more. It read:

> Here's the *YANK*, brother. ... It's not GI except in the sense that we are GI. ... Our enemies will see *YANK* as us, because it is US. ... Their psychologists will analyze the words in *YANK*, our words. What a hell of a laugh. They can't figure it out. ... Here's the *YANK*, brother.

After three issues with more such editorials, Captain Spence apologized to General Osborn that the initial tries should be considered merely experimental. His report must have been accepted, for, by the fifth issue on July 15, the ban on circulation inside the United States was lifted so far as camps were concerned. *YANK* was still forbidden to "go on newsstands or to compete with commercial magazines on sale to the public." That final prohibition never was lifted so that *YANK*, with more than 2 million in circulation, remained virtually unknown by the American public, although increasingly as the war went on it became a significant part of soldier life.

By the fifth issue, July 15, General Osborn saw something he liked, probably what he had hoped for from the time of the first White memos, a tribute to the rifle-carrying infantryman. Unsigned, like all *YANK* editorials, the memo was written by Sergeant Mack Morriss, one of the few soldiers with a military as well as a journalistic background. Editor of the Elizabethton, Tennessee, *Daily Star*, at 20 Morriss had been called into service as a member of the National Guard in 1940, and was a member of a machine-gun squad when he was reassigned to *YANK*. By then he had had two years of soldiering, had the feel for it, and, as he wrote home to his wife, was unhappy in Fort Bartholomew, with "bickering, meanness, phonies, intellectual sophistication," and a list of other negatives.

Morriss's editorial was headed "Foot Soldiers." Asked whatever had happened to the old "queen of battles," his infantry, he lamented, it's now all "glammer boys," the Air Force going "into the blue out yonder," the tank outfits "with no pack to carry," the paratroopers with "pretty silk 'chutes blossoming out," and OCS, Officer Candidate School, with "those little gold bars and the pink pants" at graduation. Morriss concluded:

> the old infantry, looked down on and despised, was feeling sorry for itself ... until one time ... a fellow spoke up for the ground pounders. ... Smoke-begrimed men, covered with the marks of battle, rise from the foxholes of Bataan, he said. Ask a man with blue braid on his cap—ask him now what outfit he's in. 'I'm in the Infantry, by God!,' he'll tell you. 'She's still Queen of Battles.'

Bataan in the Philippines had seen the Army's first combat and suffering. As a soldier, Morriss empathized with those besieged

YANK's premature call for a second front. (*Courtesy of* YANK)

troops. His editorial reflected the honest patriotism and gallantry the high-ranking founders of *YANK* had hoped would come from the EM staffers. Ironically, it was published just a few days before *YANK*'s first great crisis, a showdown on who would be in charge and what would have to come from the EM staff if the publication was to survive. An editorial on August 19, and two successive columns on "Hollywood in Wartime,"—one on the 19th and the second a week later—brought down the thunder.

Written by then First Sergeant Bill Richardson, the erstwhile *Chronicle* feature editor, the editorial had the headline "To Whom It May Concern." It concluded: "When in God's name do we fight?" The question of the moment was when the newborn American Army would open a second front in Europe, relieving pressure on the Soviets. There was no doubt about who were the intended recipients of the sergeant's open letter. They were FDR, the commander-in-chief, and Chief of the General Staff George Marshall. The two were implicitly pilloried as chicken-hearted. Writing from peaceful 42nd Street, the *YANK* sergeant insisted that, without further delay, his fellow GIs go into action.

> When do we fight? Being soldiers we have sat around for months now, waiting for that question to be answered for us. ... We came here to fight. ... We know that the world at war is a vast maze of complicated problems in logistics and transport and production and we know that fronts are hard to open. But we also know that we came into this profession of soldiering in all good faith. ... Complex be the world or not, we came here to fight. We did not come here to wait. ... When in God's name do we fight?

It was two years before D-day in Normandy. General Marshall is said to have joked about the editorial, "The only place we might be able to open a second front would be on Staten Island," a fellow New York City borough across from *YANK*'s Manhattan.

Henry McLemore, a syndicated columnist for the *New York Post*, jeered at the anonymous editorial writer. The fellow who wrote that, he said, "was sitting comfortably behind his typewriter at 205 East 42nd Street and in danger of earning his wound stripe because his portable was not resting firmly on the table. ... He showed the true public relations spirit which, in the words of Wimpy, are 'let's you and him fight.'"

Wimpy was a cartoon character of the time known to be pusillanimous.

By unhappy coincidence, it was in the same issue that *YANK* ran the first of two successive columns mocking what Hollywood was doing for the war effort.

> Hollywood, brethren, is a wonderful place. ... They come back from their camp shows ... and say 'we're doing our part. We are specialists and here is where we belong. ... We are doing as much as the boys on the fighting

fronts.' Then they go home, take a quick dip in their swimming pools and rest up for another week before the next camp show comes along.

The August 22 column mocked further. What films the studios make are escapist fluff, it said, instead of "real propaganda pictures that will inflame us to greater efforts to win the war." There was more. Usually Hollywood gets deferments from service, but when some of the actors get drafted, "they become M. P. (Motion Picture) corporals" and, soon after that, lieutenants. For good measure, there was some scorn for the diminutive, bouncy actor Mickey Rooney, and for Hollywood's dominant personality, Louis B. Mayer.

It was too much for General Osborn. As chief of "Special Services" and morale, he was counting both on Hollywood and YANK to help mould the spirits of the men preparing for combat. One of his officers, Major John B. Stanley, was already in the studios working closely with Frank Capra and Fox-Western. Much was hoped from him and them.

"I am sick at heart," Osborn told Stanley. "That and the editorial on the same day."

Something would be done about it, he continued. "White was here and I talked with him and I also called up Spence. I think that when I got through with them they knew how seriously I felt about it."

Correspondence in the YANK files show what happened next. General Osborn wrote Fred W. Beetson, the executive vice president of the Association of Motion Picture Producers, that "a certain reorganization was made in the YANK staff." With the scornful Hollywood columns in mind, he said he was sure "that there will be no recurrence of such unjust, unfair and inaccurate material."

Colonel White, the man who had dreamed up YANK, was removed from his position as officer in charge and shipped to North Africa to oversee a revival of Stars and Stripes, a service for which he later received the Order of Merit decoration.

YANK, The Army Weekly

Bill Richardson was sent off to London to develop a *YANK* edition for Britain, an assignment he carried out with such enthusiasm, but also with such persistence in imposing his own ideas of what a literate GI publication should be, that he crossed swords with the new administration on 42nd Street, volunteered to quit *YANK*, and found his resignation accepted. He was transferred to other Army work.

About the same time, Captain Spence, the anomalous officer editor of the GI publication, was shifted to other military duty. While the brass now made clear that there were limits to *YANK's* freedom, they were still willing to give a GI-run paper a try. In Richardson's successor they found the GI editor they were seeking.

YANK

Turn 9 Around

YANK's new managing editor was Joe McCarthy, an Irish American sportswriter from Boston, a bulky man's man who came to the magazine during its first month of existence wearing the two corporal stripes he had earned in the mule-mounted 97th Field Artillery.

Joe shared with his predecessor a sympathy for his fellow enlisted men, but he had no antipathy for the officer corps, unlike the hints in an August 12, 1942, editorial, "Army bosses," published in the Richardson era. That editorial had expressed pleasure that "a breakdown in the old heel-clicking" seemed to be under way. While conceding that there was a need for "discipline and the power of command," the editorial called for "more 'skippers' and fewer 'commanding officers.'"

In one of his scores of round-robin letters to YANK's correspondents abroad, McCarthy explained his differences with his predecessor and with Richardson's handling of the London edition.

"Personally," he said, "I am getting fed up with that old worn-out line about standing up for the cause of the enlisted man. Mentioning no names, we've had more than a couple of people on YANK who have gotten into jams of their own making and then have tried to claim that it happened because they were making martyrs of themselves on behalf of the enlisted man's rights."

"In most of these cases," the erstwhile mule skinner went on, "the so-called martyrs wouldn't know a real enlisted man if they fell over one."

A crisis, McCarthy said, had developed between the New York master edition and the British version because of a basic disagreement on what a soldier publication should be. The YANK rule was that local editions, all of them staffed by EM, could drop one-sixth or so of the New York copy, replacing it with local news, but the British version had insisted on much more space of its own. In the New York master edition, McCarthy said a GI flavor had been developed so successfully after months of trial and error that other branches of the morale division, such as those in radio and motion pictures, unable to win a similar degree of EM confidence and good will, "are coming to us to ask us to help them be GI." He went on, "one outfit wanted to station one of their lieutenants in our main office 'to soak up some of YANK's GI tone and flavor.' ... We didn't bother explaining that you cannot get a GI flavor that way. YANK has it because the men who are writing our stuff are (now) living with the Army day in and day out and because the men who are putting that stuff into the magazine are enlisted men not lieutenants who have never drunk out of a canteen or who, like one we heard about in the Radio Section, drew a red pencil through such expressions as "sweating it out' because 'it sounds vulgar to me.'"

YANK had a side office for the editor, but McCarthy made no use of it, preferring instead to stay in the main room along with the rewrite men and other staffers. Richardson in London, by contrast, he heard, had a separate space shared with his assistant managing editor while the rest of the staff worked in one big room. The disaffected lowlier editorial troops, McCarthy heard, referred to the managing editors' office as "the House of Lords" and their own as "the House of Commons."

Richardson's removal, McCarthy told his far-flung staff, was not caused just by differences in managerial manners and the extent of

YANK world headquarters on 42nd Street, "Fort Bartholomew." Joe McCarthy, with his six stripes, is seated at the central table at the right front reading copy. (*Courtesy of* YANK)

the scuttling of master edition material, but also because of the very nature of the substituted copy.

"Bill," said McCarthy, "favored the ultra sophisticated kind of humor and the melodramatic, stylized kind of action stories which usually opened with a terse statement like 'this, then, was it. This was what they were waiting and training for all these months.'" In the London editor's view, McCarthy added, "every page should sing (and) most of the pages he edited did, unquestionably, sing, but they were always rewritten and hyped so much that they bore little resemblance to reality. And they were certainly not GI as *YANK* has to be."

The crucial turning point for *YANK* came with sportswriter McCarthy's quick early jump from corporal to six-stripe master sergeant and his transfer from *YANK*'s sports beat to general editor. It came at the significant time in late 1942 when the Army went into action in North Africa and 42nd Street soldiers could be shipped out to cover the fighting.

Before that, the military had the same trouble converting *YANK*'s ex-civilians into soldiers as all other parts of the service had with their recruits. In *YANK*'s case, the problems had been especially bizarre.

To fit into the military pattern, *YANK* was organized as a company, a military unit larger than a platoon but smaller than a regiment. A service-minded Marine was brought aboard as a first sergeant. Soon, once each week, he had the cartoonists, photographers, sketch artists, editors, writers, copy readers, rewrite men, leg men, accountants, and circulation and promotion specialists taking off an hour or so from their publishing efforts to march to his "hut, hut" in close order drill on a field near the East River end of 42nd Street or in Central Park. Sometimes the staffers were ordered to play baseball, a way to get exercise. There were no DiMaggios, and some just fanned the air with bats as pitched balls flew past them, but play they did. By October 1943, there was a new wrinkle. From 8:00 to 8:30 each morning, before going to their typewriters, photography darkrooms, and artist easels, the soldiers of Fort Batholomew had to do calisthenics. There was patient compliance, but as McCarthy murmured at the time, "it makes a long day."

As in any military company, *YANK*'s Marine first sergeant maintained a sick book and a duty roster and he followed normal practice by making out a morning report. The difference was that the bulk of *YANK*'s 150 staffers soon were scattered across the globe in more than 20 countries. To keep the Marine first sergeant and Joe McCarthy in touch, the faraway staffers were told to send back each week something to print, or at least a letter telling what they were up to. As for getting exercise, each one was on his own.

Usually it all worked, but at one point McCarthy complained in

Joe McCarthy. (*Courtesy of* YANK)

his letters to the field that there were some abroad who "let three or four months slide by without ever making the slightest gesture about turning in a story or even writing a letter to give us a vague hint about whether or not they are still in the Army." The GI editor listed 10 who kept in touch, but he said, "they are about the only ones who take the trouble to either send us stuff regularly or keep us posted on their activities and plans so that we will have some idea of what the score is. How in the name of Christ do you expect us to run this organization intelligently otherwise, I don't know."

To keep such an extraordinarily diffused military company together, McCarthy did his part with his periodic letters, often sending them on a weekly basis. He filled everyone in on what others were doing, praising many and cracking the whip sparingly, but effectively, when necessary. With tongue in cheek, in letter number 47, on February 14, 1944, he took note of the letters as a *YANK* institution by themselves.

"By the way," he wrote, "we resent the way so many of you people scornfully refer to these priceless bits of mimeographed correspondence as 'poop sheets.' Don't you know they will be collectors' items one day?"

That "one day" seemed far away in the first weeks of 1944, but McCarthy was right. The 67 in the series now offer unique insights into how the once uncertain *YANK* evolved into the real voice of the GI it was intended to be.

YANK
Smell 10 the Japs

In the dozens of letters the New York home office sent to its correspondents across the globe, one event recounted in them illustrated the marked difference between *YANK*'s faltering first efforts and the no-nonsense depiction of soldier life that the magazine became.

The first of the round-robins were written by Captain Spence, the minister's son who was the first editor. Brand new to the Army, he tried to think in the mind-set of the way a soldier would think. In the initial letter of the series, on November 19, 1942, the captain instructed distant staffers not to limit themselves to entertainment features but, as often as possible, to get copy on combat.

"We need a little more hair on our chest," was the way he expressed it.

When Sergeant Mack Morriss, in the Solomon Islands, sent in some grisly accounts of battles, Spence, by then a major, seized on them to get Sergeant Merle Miller, in Hawaii, to emulate the example. Merle had a column in the Hawaii edition chronicling the fictional adventures of an effete Dudley Wintergreen, a fellow with more of a taste for the finer things of life than for a dogface's drudgery.

"You can actually smell the Japs in that Morriss piece," the major exulted. "That's the kind of thing we want for *YANK*."

The major was, indeed, on the right track in seeking combat coverage, but the reality of often dreary rear echelon soldier life, summed up as "hurry up and wait," provided Miller with no way to oblige. As McCarthy noted in a later round-robin, "Miller was stationed in Honolulu where there were no foxholes and there wasn't anything he could do about it." McCarthy told how Miller replied, "The only Japs, sir, that you can smell in Honolulu are those that ride in the electric busses. This isn't a combat zone."

Still, raids occurred on western islands, so Miller left his pleasant Waikiki bungalow to cover the costly Kwajalein invasion. Troops back at Fort Bartholomew got a laugh out of the needling lead on Miller's report, "The smell of dead Japs is everywhere on this island tonight."

With the soldier world so far removed from the delicate atmosphere of the church rectories in which he had been raised, Major Spence tried in each letter to couple good writing advice—get lots of soldier names and hold down on use of the first person singular—with what he deemed an appropriately lewd barracks-type closing joke. A typically tasteless example, on December 4, 1942, concerned a sailor who called his waterfront pickup "Hollyhocks," because the dictionary said that such was "no good in beds" and that the "best results were against sides of buildings and fences."

By contrast, the erstwhile pilot of Army mules, McCarthy, rarely attempted extraneous humor, limiting himself to directives to the staffers and office gossip. An exception was at his wife's expense on February 14, 1944. Charlie Chaplin, the comedian whose films were early Hollywood classics, had been charged in a sex scandal. McCarthy concluded his 1,500-word round-robin, "my wife is getting so that she thinks in Army terms. A few days ago Charlie Chaplin got himself socked with a federal indictment in the Joan Barry case. When my wife saw the headline in the papers, 'Chaplin and six others indicted on girl charges,' she shook her head and said, 'my goodness, I hope it isn't a Catholic chaplain.'"

McCarthy was determined that *YANK* be an honest expression of soldier life and needs. To that end, he used the circular letters as an editorial conference with staffers on distant continents sitting in on his discussions of what the magazine should do. On April 27, 1943, he took up a burning soldier topic, troop rotation, especially where it concerned EM in the sweltering tropics:

> We have been trying hard to get the War Department to state a definite policy on the problem of relieving men who have been in the tropics longer than 18 months. We have finally gotten a definite policy but it isn't any too cheerful. Their argument is one of transportation. They have a hard enough time supplying transports to places where there are active forces without getting the necessary boats to send out a task force just to relieve another task force where there is no fighting.

The WD had a fairly good argument, McCarthy admitted, although "we're not giving up on it yet." Meanwhile, he urged, "Kill any rumors in your area about everybody going back to the States after any definite period of overseas service. Those rumors build up to too much of a letdown."

On May 3, 1943, McCarthy disagreed with Miller in Hawaii, who thought that a superb April 23 account of C Company's January mop up of Japanese resistance on Guadalcanal was "great," but lacking "something in timeliness." Written by Morriss and illustrated by Sergeant Howard Brodie, it included soldier-by-soldier comments about face-to-face combat. Said one soldier, "There were plenty of snipers but y'know them damn guys couldn't hit a bull on the back with a shovel full of sand."

Too late for publication? Not according to McCarthy's idea of what *YANK* should print.

"Maybe we're wrong," he said in the name of the home staff, "but we don't look at it that way. This is a magazine with very little pretense about being up to the minute and isn't a good story like that piece still readable even a year or two after the battle described?

That doesn't apply to every story, of course, but it does apply to a really good piece of writing."

On May 23, 1943, the round-robin had a chitchat about the long, slow sea voyages some staffers took on the way to assignments at a time when intensively revved up war production had not yet produced enough planes. Making out the military travel orders was the task of Annie Davis, the 20-year-old civilian secretary and most ornamental occupant of the home office's editorial room. Trying for fast transportation though she did, she was frequently unsuccessful with the result that one lamentation came from Sergeant Walter Bernstein. He and Sergeant Marion Hargrove were outbound on a slow transport, Bernstein heading for Cairo and Hargrove going to China.

The round-robin shared Bernstein's message, "Give my regards to Annie Davis who arranged this trip. She should live so. Every other day we dive overboard and pace the ship for a while. Nothing happens except some birds that fly alongside and make faces at us but Hargrove made faces back at them the other day and they haven't been seen since."

A recurring theme in what amounted to an editorial conference by mail was just how free were *YANK*'s EM to publish anything they pleased. After all, *YANK* had officers and the military had censors.

McCarthy addressed the first on October 18, 1943. Major Spence had been transferred off *YANK* to other duties, so no officer any longer served as executive editor. The decisions on magazine contents were left to McCarthy and his EM counselors. "Colonel Forsberg," he said, "checks page proofs for pornography but no one else ever sees the material in an issue before publication." "It certainly doesn't seem," he added, "as if the brass hats were dictating the policy of this paper."

In other letters, the sergeant conceded that *YANK*'s staff had to exercise its freedom within limits. One such incident concerned politics. When *YANK*'s Italian edition asked for an interpretative column on events back home, Sergeant Miller, back in New York after 18 months in the Pacific, was assigned to write it. It was the summer of 1944, just before FDR ran for a fourth term, and McCarthy sympathized with Miller's task:

> (The column) is a bitch to write when it concerns politics, and politics, of course, is the big item in home news these days. In order to give a reaction … you have to set yourself up as a voice of American opinion. But if you say, for instance, "everybody at home thinks that (Thomas E.) Dewey is a lead pipe cinch to win the Republican nomination," you violate the War Department policy which says that you can't give such opinion unless you quote a wire service or some recognized authority. How the hell, then, can you write a decent interpretive letter under those conditions.

The sometimes-capricious censorship in the field was the second problem with many discussions of it to and from those abroad.

In one of his final efforts in Hawaii, before reassignment to New York, Miller had provided McCarthy's round-robin of December 14, 1943, with an example of how censorship and other problems made it a challenge to get out overseas editions. An invasion was underway at the atoll of Makin in the Gilbert Islands. For *YANK*, Corporal Larry McManus was there as the writer and Sergeant Johnnie Bushemi the cameraman. *YANK*'s copy was supposed to be more or less timeless, but Miller still wanted the report to be as up-to-the-minute as possible.

Normally, *YANK*'s Hawaii edition went to bed at the printer's on Friday, but Miller held it open hour after hour until, on Monday, the *Honolulu Advertiser* paper bawled him out for delaying their own press run. *YANK* used all *Advertiser* facilities.

With the issue finally on its way on Monday, Miller reported to New York:

the book went to press this afternoon and just how I don't know. ... At 6 P.M. the book is run off and I come home and I am now tired and I wish to hell I'd decided to be an infantry private in this goddam war. ... I was holding open the damn thing until Saturday, thinking we'd get copy from McManus and pictures from Johnnie but no. No copy. Everybody else gets copy but not us. ... I put the thing to press at the last possible minute Saturday afternoon.

For the Makin story, Miller had decided early Saturday to bury it on pages six and seven, rewriting civilian press dispatches. He did not "want to play up that phony stuff in the front of the book." For "art," he chose a map and "some lousy training pictures of the 27th Division." Then came stop-press developments. Some of the Makin wounded arrived in Honolulu at 3 P.M. Within an hour, Miller received authorization to interview them, and by 5 P.M., had "a fair yarn from a sergeant." At 6 P.M., Bushemi walked in. He had photos of the Tarawa landings but, as for Makin, he said, he had sent them on to *YANK* a week earlier. That called for an encounter with the censor.

"John and I whip out to the Navy," the unstrung editor reported. "The Navy censor maintains a bored indifference toward the press services and the same nonchalant attitude toward *YANK* which, since we are a weekly, he can't see any use hurrying about it. Finally he consents to get two of John's pictures passed by Monday morning, a day after the last possible deadline. I throw away the (two pages of press copy rewrite) at 7 P.M. and ... write the story of the sergeant."

At 8 P.M., to illustrate the sergeant story, Bob Greenhalgh, then on the Honolulu staff, rushed to the hospital to sketch the wounded soldier. Photos were banned. At 9 P.M. he was back, but G-2, the intelligence branch, killed his picture. The non-com had been sketched in bed. No portrayals of beds, said G-2.

"Greenhalgh in suicide-contemplating disgust," said Miller, then "draws a sketch of what he thinks may be a likeness of the sergeant's head. It is. G-2, of course, says okay."

Smell the Japs

Not that much in a hurry and more interested in lasting prose, the master edition, weeks later on December 31, gave the first four pages to McManus's "Shamrocks at Makin," an eyewitness account of how a battalion of New York City's historic "Fighting 69th" regiment wiped out resistance on Butaritari Island in the Makin atoll in a "wild mad night ... (amid) a bedlam of infiltration, screams, laughing, and suicidal charges against the American defense perimeter." Bushemi contributed seven photos, including a full-page cover shot of two GI riflemen firing across a Japanese corpse.

Although censorship of the Makin battle proved mainly to be a question of delay rather than obliteration, the round-robins included worse cases of the killing of editorial material. On July 31, 1943, McCarthy told of what happened to Sergeant David Richardson, a one-man writer-cameraman "team" in Southeast Asia. "Dave," reported McCarthy, "didn't do so well photographically in the expedition to Woodlark. He took back 23 pictures and the censor killed 20 of them. Evidently they killed a couple of his stories, too, because we received only one from him."

In May and June of 1943, managing editor McCarthy offered counsel on what to do and not do with regard to censorship. Sergeant Al Hine, editor of the Teheran edition, helped him with the first:

Al writes from Iran requesting an editorial campaign against the practice of labeling certain good cafes in foreign towns 'in bounds for Officers only' or 'in bounds for Officers, Warrant Officers and Sergeants only,' which knocks the hell out of the average dogface who just came from the U.S. hearing a lot of stuff about fighting for the Four Freedoms.

Al points out that he is including the idea (which we will use, of course) in a personal letter rather than writing the editorial himself because he thinks that such a piece written for publication might not get through his local censor. That is something for the rest of you to remember. You can often get the local gripes to us in letters that wouldn't pass as stories intended for publication.

101

What not to do with regard to censorship was spelled out in exasperated fashion later, on June 9, 1943, when McCarthy begged the correspondents in the field to show some prudence and to accept the reality that they were enlisted soldiers inside the Army. With anguished emphasis, using all capitals for some words, he wrote, "if we've said it before, we've said it a hundred times, but it still doesn't do much good. FOR GOD'S SAKE, DON'T fill up your letters to the office with complaints about the censors and the lack of cooperation that you are getting in your command. Whenever a correspondent does that it gets us in wrong with the local command immediately and, at the moment, we are in wrong with at least two commands for that reason. We don't mean that you should never kick about local conditions but please try to remember that your mail is being read by the very same officers that you are complaining about. If you continue to call certain second looeys sons of bitches you will only make it very tough for yourself and us, too."

The ex-nursemaid of artillery mules broke into school Latin and a tangle of metaphors as he urged his readers to heed the words of wisdom: "*Verbum sapientiae sat,*" he concluded, "and, if the cap fits, put it back on the rack where it belongs because the owner may have dandruff."

YANK

Out of 11 the Office

Many things went into making *YANK* the faithful voice of the World War II serviceman.

The man in uniform often was lonely, thinking of the wife or girlfriend at home. The weekly pinup of a beautiful woman was a stand in.

Sergeant Baker's comic strip of the Sad Sack, the most hapless private in the Army never elevated to private first class, was a comfort to EM. Misery loves company and, no matter what frustrations a soldier had, there was satisfaction in contemplating someone, albeit fictional, who was worse off.

Other cartoons by staffers and volunteers in service poking fun at lowly second lieutenants, or even at generals, were additional sun rays on days of rain.

The "Mail Call" feature allowed everyone among the 14 million in service to raise questions and air grievances, not just inside a hut with an audience of 10, but with millions on all the continents. Officers, as well, needed somewhere to sound off, so *YANK* lifted the usual ban to give them, too, access to "Mail Call."

Each feature contributed to *YANK*'s rapid and steadily increasing success but, most important of all, the early decision was the recognition that a *YANK* confined to Manhattan's 42nd Street would never be GI and could never understand and speak for fellow inductees.

103

A decision was taken early that all but a handful of key specialists, such as managing editor McCarthy and Sergeants Weithas and Ralph Stein in the art department, soon would have to get out of Fort Bartholomew, going abroad for at least a half year. Even those few were put in question in the spring of 1944, when the War Department set July 1 as the date to ship abroad anyone in uniform who had been 12 months in the United States. Most staffers, such as Sergeant Greenhalgh, took for granted that they were in service to cover the war overseas, but some had no objections to staying in Manhattan. McCarthy discussed it in his March 31, 1944, round-robin:

> Great topic of conversation here in New York these days is the new War Department policy. ... It may be that *YANK* men will not be exempt from this ruling even though we revolve our own men and even though we are specialists who would be hard to replace overnight so a number of the less adventurous characters who are not keen on traveling are understandably perturbed by the whole thing.

In that regard, McCarthy poked some fun at Stein who was a prolific and gifted humor cartoonist:

> Last week somebody said to Stein "would you care to make a statement on the new troop rotation policy, sergeant?" Ralph turned his billiard ball eyes up to the ceiling and chewed on his paintbrush "I am turning it over in my mind," he announced. "Or, perhaps, I should say, I am turning it over in the pit of my stomach."

Sergeant Mack Morriss, who had written the paean to the queen of battle, the footsore infantry, was one who had no problem leaving, and among the first to go. Four months after the birth of *YANK,* in October 1942, Morriss set out for Guadalcanal where Marines were giving the Japanese ground forces one of their first setbacks.

Mack admitted to mixed feelings. In letters to his wife and in memos tracked down by Professor Ronnie Day, chairman of the history department of East Tennessee State University, Morriss said at departure that he "felt at once a great sense of relief and a great dread."

The relief was in getting out of "a center … of military folly, of political meddling, of waste and indecision, … of the false patriotism of War bond cuties, of bad movies and of stinking radio plugs." The ex–machine gunner could not have been more scathing in regard to *YANK*'s faltering first efforts and to the stay-at-home's view of the war, but Mack had several concerns.

"I dread the waiting." Everything in the field was "hurry up and wait." The second was made up of "two sub-fears, that I will be where there is no danger (or) that I will find more than I can survive."

With Morriss as illustrator went Private Brodie, a sports artist for the *San Francisco Chronicle*. *YANK*'s early managing editor, Richardson, had known him as a fellow staffer and had recruited him for the magazine. The Morriss-Brodie pair were to become two of *YANK*'s finest teams—Morriss a deftly sensitive reporter of soldier anguish and bravery, Brodie a daring recorder of battlefront horrors. Their experiences were typical of what other *YANK* writers and illustrators would experience after them. The two sailed west across the Pacific to Auckland, New Zealand, on a troopship, the converted Dutch liner *Tjisadane*. To keep busy on the long passage, they put out the ship's illustrated newspaper, the *Salt Water Taffy*.

Then Major Forsberg gave the two an introductory letter to the commanding general of USAFISPA, the U.S. forces in the South Pacific area, headquartered in Auckland. In effect, he asked that the sergeant and private be allowed to wander freely, see combat as they chose, and be exempted from any other soldier chores.

Lieutenant Sam Humphus, the assistant executive officer of the Amy's Special Services division of which *YANK* was a part, provided each with additional paper, "As a representative of *YANK*, you are a marked man. As newspapermen, you are the custodians of the high traditions of honor, reliability, and aggressiveness for which American newspapermen are distinguished. As soldiers, you are representatives of enlisted men of the United States Army. In discipline, military courtesy, conduct, courage, the utmost is expected of you."

Ten weeks after getting the Humphus orders, the two, wearing the twin hats of soldier and chronicler, climbed down from a B-17 bomber onto the metal plankings of the newly built Henderson airfield on Guadalcanal. It was Christmas Eve 1942, and they were just in time for the clean up of Japanese resistance.

While each wore the headpiece of newsman, the two discovered promptly that the hat of GI seemed to be the one on top. They were unwelcome to move into the press center where Henry Keyes of the *Express* of London and American reporters and photographers of AP, UP, Hearst's INS, and Chicago's *Tribune* and *Sun* were bedded down. Making the best of it, they put up their own pyramidal tent, labeling it in un-GI fashion the Artists and Writers Club, Cactus Heights, Bomb Alley, Guadalcanal. Cactus was the island's code name.

When bombs fell, the *YANK* pioneers were allowed into the "Cave," the shelter used by Keyes and the other civilians, but when it came to mealtimes the others, with their simulated rank of major, dined in the officers' mess. The two from *YANK* walked in the EM chow line. When the civilians got whiskey, the two from Fort Bartholomew sipped canteen water.

Most frustrating was when the journalists raced off in the single press jeep to cover a story. The civilians climbed in first. When no room was left, the *YANK* pair walked.

Morriss had lost his typewriter so he borrowed one when he could.

When "Cactus" petered out, Morriss summed up his reactions in a report to the home office. The civilians of the Fourth Estate were a "swell bunch of guys," he said, but "this may sound silly ... I felt like a bastard child the whole time because I had to depend on somebody else to supply me with the essentials of this racket."

Morriss's and Brodie's eyewitness reports of soldier life in the Solomon chain were accurate descriptions of both the danger and the boredom, but the New York staff, still new to war coverage, felt the need to hype and glamorize. On March 5, 1943, a line along the bottom of page one announced, "Exclusive battle sketches from Guadalcanal, pages 2–3." Six Brodie sketches and a Morriss piece called "Jap Trap" filled the second and third pages and one-third of page four. A box boasted, "These pictures were sketched under fire." Morriss was furious when he saw a copy in New Caledonia. Brodie, like many combat artists, took scribbled, and sometimes just mental, notes under sniper fire but usually completed the sketch later in the rear.

Morriss had called his straightforward account of combat simply "The Story of a Battle." He quoted one infantryman, for example, "Them little bastards won't give up—they're too damn ignorant."

"Jap Trap," he protested. "How cute." He went on angrily, "Somebody fouled up Brodie's stuff. The box says 'These pictures were drawn under fire.' ... Two of them were not and it was obvious they weren't, and you must have known they weren't, because Brodie always sent a note with each set of sketches explaining where they came from and why." One was a placid scene of a tented camp area. Another showed the evacuation of the wounded. As for Morriss's own story, "Tell the copy desk they have my compliments and thanks for a beautiful job of butchering. They didn't miss the point, they messed up the whole goddam story in the details. The hell with it. I'm disgusted."

Mack got over it. He covered the battle of the island of New Georgia, which lies between Guadalcanal and Bougainville. When both he and Brodie caught malaria, the two were rotated home for a while to work stateside. Then both went to Europe to cover a very different kind of tank and artillery fighting.

In November 1944, on the fourth anniversary of the call-up of the Tennessee National Guard, Morriss tracked down his original company, a unit of the 30th (Old Hickory) Division, which was advancing through Maastricht in Holland toward Germany. Mack filled two pages with the story of "My Old Outfit." This time rewrite made no effort to jazz up his poignant prose.

Of the original 150 in his old Company A, only four men were left. Some had gone on to OCS to become lieutenants. Some were dead. Harry Nave, the company clerk, had tried out for the Air Force and was killed in training. Clyde Angell, who "talked with the nasal twang of East Tennessee," had reveled in a deep log-covered hole in Normandy but "Jerry (the Germans) came over, dropping big-stuff bombs straddling the shelter," and A Company's kitchen "was blown all to hell." A severe concussion killed Angell. Herman O. Parker was still driving a truck. He and Mack watched an infantry patrol coming back from the front. "One of us?" asked Mack. The answer finished Mack's piece. "Doggone," said Herman. "I don't know. I don't know anybody in the company any more."

Morriss's reports and Brodie's sketches set a standard soon followed by other staffers as *YANK*'s word and picture coverage was soon recognized both inside and outside the Army as among the war's best. There was no longer any worry about the magazine's credibility. With hundreds of thousands of copies of each issue in circulation, it was estimated that there was an average of two or three readers for each.

YANK
The 12 Sack

If, half a century later, the present generation believes that the World War II soldier was much different from his grandson, that he was happy to face the challenges of combat and stoically receptive to the thought that he might soon lie beneath a cross or star of David, that illusion is dispelled by a glance at one of *YANK*'s cornerstone features, the Sad Sack comic strip.

It was the work of four-stripe Sergeant George Baker, a prewar animator for Mickey Mouse and other Disney productions. His Sack was cathartic, reassuring the most miserable of privates that at least one imaginary wearer of the uniform was worse off than he. Further, though exaggerated, the woes of the Sad Sack struck millions of enlisted men as not that far from the truth.

Although it never appeared in print in *YANK*, the Sack got his name from a foul expression familiar to all in service. Someone completely out of touch, a loser, was, alliteratively, "a sad sack of shit."

The Sack suffered every humiliation and injustice, usually at the hands of sergeants, officers, and generals, the very ones whose responsibility was to bolster his morale and win his trust. Happily, instead of provoking rebellion, the Baker strip metamorphosed grievances into laughter and eased the pain it depicted.

Some military units even identified themselves as a collection of Sad Sacks. An example were 67 fighter pilots in Tunisia in the spring of 1943. They had been trained in the limited techniques of low-level strafing and had been shipped abroad as sergeants, not as the usual pilot second lieutenants. They wore an insignia of their own designing, a round patch with a huge 67 and the Sack name below it. The raids they conducted helped drive the Axis out of Africa. For *YANK*'s edition of June 11, 1943, writer/photographer Sergeant Pete Paris produced a cover story about the short-changed 67.

The struggle for promotion inside the military was universal, a theme to which Baker constantly returned, while remaining himself a staff sergeant, the fourth level from the bottom in the military pecking order. On December 9, 1942, the strip recorded one of the Sack's first efforts to leave the basement level of service. He noticed that big bellies seemed to go with the six-striped master sergeants and the five-striped first sergeants, so he tucked a pillow under his shirt and requested a promotion. For an answer, his lieutenant kicked him into the street.

For the February 2, 1945, edition, Baker had the Sack thinking that his job running his unit's travel section might call for the single stripe of pfc (private first class). Again, no luck. Not only did he lose his comfortable job, he saw it turned over to two of his military betters, Lieutenant Pacrat and his aide, the four-striped Sergeant Tubb.

A month later, the Sack did such a fine job building a road by breaking rocks and hauling timber that it was clear that somebody should be rewarded with a promotion. The Sack's sleeve stayed virginally nude while a colonel raised a deskbound lieutenant to captain.

Proof that the readers were sweating out the Sack's efforts to advance was spelled out in letters to the magazine. In the July 9, 1943, issue, five sergeants of Camp Swift, Texas, with 20 stripes among them, appealed for a pfc stripe for Baker's big-nosed little man. "He's not the most brilliant soldier we've seen," they conceded, "but he tries to the best of his ability."

A Sgt. George Baker "Sad Sack" cartoon. *(Courtesy of YANK)*

Nothing doing, was the magazine's reply, along with a hint of *YANK*'s own lamentations about rare promotions, "Ratings are pretty damned scarce around here."

Baker teased the Sack's fans on March 5, 1943, by showing the pen-and-ink character sewing on the two stripes of corporal. False alarm. In exchange for a few coins, the Sack was just manipulating needle and thread on behalf of a friend.

The Sack never rose even to pfc but, for a moment in the summer of 1945, it looked as if he had his chance. A new Army Regulation, number 615-5, amended the tables of organization to allow an unlimited number of pfcs while retaining limitations on higher grades. The Sack was in? No. His unfeeling commander pointed to the fine print. The promotion was in order, but only if the candidate was "qualified."

Baker's strip appeared in the first dummy edition of June 6, 1942, and in each issue thereafter. The cartoonist was discharged from the Army soon after the Japanese surrender, but he left behind enough cartoons to reach to the end on December 28, 1945.

Tirelessly inventive, Baker poked fun at every aspect of Army life, beginning with the physical examination of inductees in issue number one and continuing until the Sack's discharge in YANK's final issue. The first strip showed a smirking mailman handing the week-kneed, thin-chested, bulbous-nosed Sack an envelope marked "War Dept." Although men in their forties and fathers of families were being scooped into the troop-hungry Army, the frail Sack was a prime candidate for the 4F category of the physically unfit. Pushed around by medics twice his size, and limp from intrusive examinations, he was dragged away in the last frame, his back adorned with a huge 1A proclaiming him militarily competent.

The end of the series, entitled "Happy Day," portrayed a jubilant civilian-dressed Sack out of the Army. His joy was brief. Radio sets and newspaper headlines blared the news of the world to which he was returning, housing suffering from the worst shortage in 10 years, inflation spreading, a postwar diplomatic crisis looming, 3,000,000 jobless foreseen for the spring of 1946, and fears that an atomic rocket could wipe out America in 30 minutes. The final panel showed the despondent, newly hatched veteran slumped on a curbstone.

Between the bookend cartoons of 1942 and 1945, Baker limned more than 150 ways a private soldier could be unhappy, first in basic training, later in combat.

In 1942 and 1943, as the draft and a booming war industry prepared to engage the enemy, the Sack's woes were home-based—close order drills with the Sack poked in the eye when all swung right and he twisted left, inspections when his best effort at neatness ended in a gig for a petty slipup, and paper-cluttered camp billboards with one at the bottom from December 25, 1776, reading, "notice, all men will fall out at 4.00 this morning and proceed across the Delaware, by order of General George Washington." Further, there were films warning about venereal disease in such frightening fashion that the Sack used a rubber glove before shaking hands with his buddy's girlfriend.

Next, there were gold-brick slackers getting sergeant stripes while the eager beaver Sack got none, lieutenants stealing away his girl-friends, generals treating him as a lackey, and "officers only" signs in front of good hotels and the sole "enlisted men only" placard posted on a latrine.

With the foe engaged in 1944 and 1945 in North Africa, Europe, South Asia, and the Pacific Islands, the Sack had new concerns: quicker-footed comrades jumping into his foxhole exposing him to falling bombs, and the Sack hurling grenades in the drive toward Paris only to come up against a Porte d'Orleans sign reading, "Off limits to American troops." Boasting placards, such as "You are now entering Germany, courtesy 79th Inf. Div.," were being posted so Baker provided one for his character, a freshly dug latrine proclaiming, "You are now entering these premises through the courtesy of Pvt. Sad Sack." At a pay of $60 a month, the infantryman Sack cleared the way for a 60,000-ton tank. As a veteran of the ETO, the European Theater of Operations, he headed home to the United States with joy only to be sent on farther to fight in the Pacific. Shipwrecked en route and waterlogged, he just managed to swim toward shore only to come up against a sign: "Beach reserved for officers only." In the December 14, 1945, issue, Baker had the Sack accepting Japanese bows graciously, only to be assaulted with the usual verbal abuse when he stepped back into camp.

The requirement that lower ranks salute their Army betters as a mark of service unity grated on many in uniform, especially when a 40-year-old private might encounter a 22-year-old lieutenant graduate of ROTC or of OCS. On *YANK* itself, one college-graduate corporal on duty in Washington made a practice of jaywalking in mid-block rather than raise his arm in salute.

Mandatory saluting was a theme to which Baker returned repeatedly, causing at one time a flurry of letters. On February 10, 1943, in a strip entitled "Rank," the half-asleep Sack missed saluting a lieutenant who bawled him out so furiously that the one-barred officer failed in his turn to see and salute a double-barred captain. He, similarly indignant, missed a major, he a full eagle colonel, and he a one-starred brigadier general. Each explained himself passing the blame down to the Sack at the bottom.

Returning to the theme on July 14, 1944, in a strip entitled "Double Trouble," Baker touched off a lively discussion. Busy digging his foxhole, the Sack failed to notice a snaggletoothed captain. Reprimanded for not saluting, the Sack made up his mind not to repeat the offense. His shovel on the ground, he snapped off smart greetings to a long line of lieutenants only to bring down a sergeant's wrath for not finishing the hole.

That brought mail to YANK. Both Corporal C. Grablewski, in Camp McCain, Mississippi, and Lieutenant Norman Lipkind, at Fort Ord, California, were among those writing in. "Since when does a man on detail have to salute an officer?" the corporal wanted to know. YANK admitted that "the captain had no right to issue the order" but added that "even the Sad Sack knows better than to disobey an officer." Lieutenant Lipkind, for his part, was incensed. Field Manual 21-50, he pointed out, specifies that "military courtesy and discipline" does not apply when a soldier is at work. He went on, "Sergeant George Baker evidently does not remember his basic training very well. He has loitered too long in the environs of YANK. Perhaps a little field work would do him some good."

Even though the officer himself was stateside, "field work" could be taken as a euphemism for combat and a veiled insult to a soldier at home. EM fans, and some officers, jumped to Baker's defense. Private Lew Gulden, in Camp Maxie, Texas, wrote, "many times on maneuvers we got a nice chewing out for not coming to attention and saluting when digging our foxholes. I know just how the Sad Sack feels."

Private Adrian Robbins, at Lowry Field, Colorado, said he had been gigged once for not saluting while on detail and that by a lieutenant "right out of OCS," who likely knew no better. Still, he insisted, "it proved Sgt. Baker was right."

Camp Maxey generated another letter, this from a private, a corporal, and three sergeants. "In some parts of this camp," they wrote, "each man on detail salutes. It's an order."

Most exercised of all was one of Lipkind's fellow lieutenants, A. V. Burns, of Macon, Georgia, who protested, "The so-called environs of *YANK* include a lot of bloody, battle-scarred bits of earth that have never been mentioned in a field manual. *YANK* is being capably handled by GIs and so is the winning of the war. So let's both get the hell out of the picture and leave the guys alone."

Baker went back to the question of salutes twice more. On December 1, 1944, he sketched "Double Trouble 2," a lieutenant protesting the lack of a salute stateside and then, in combat, objecting to it because it might make him a sniper target. Strip "Double Trouble 3," on February 23, 1945, was a variant on the theme, the Sack saluting smartly both in a stateside camp and then later in an "overseas hell hole." The trouble this time was that the lieutenant nailed him at home for sloppy attire, while he objected under fire that the Sack was turned out too neatly. He was a morale threat, for fellow soldiers might think that such a paragon of dress must not be doing his military part.

YANK's readers kept a sharp eye on the sergeant's product, catching him in any error. Sergeant J. J. Gibbs, at the Willow Run Airport in Michigan, noticed on April 28, 1944, that a reclining Sack had big toes on the left side of each foot. *YANK* ran interference again for Baker, challenging the critic "since when did two left feet keep anyone out of the Army?"

In his running criticism of the Army seen from below stairs, Baker did slip on occasion. When he shipped the Sack from the ETO to the Pacific Islands he ignored an Eisenhower order of June 5,

1945, that no soldier who had been in combat in Africa or Europe would be sent to the war against Japan, even if he lacked enough service points to merit discharge. That should have given the Sack a double out, for by Baker's own count his little man had accumulated 30 points more than was necessary to leave service. He had 48 for that many months overseas, 54 for four and one-half years in uniform, and 30 for battles in which he had fought. Still, the Sack could not get out. The Army classified his job cleaning stables as "essential."

Just before the D-day invasion of France, *YANK*'s London editor told New York that in Britain the Sack was among the magazine's most popular features. The same was true all over the world, Master Sergeant McCarthy replied.

On October 29, 1943, Pfc. Robert W. Ford, on Third Army maneuvers in Louisiana, commented that the Sack strip "is not only a great deal of laughs but it's very true of Army life." He asked for Baker's photo. *YANK* responded with a gag sketch, giving Baker his creature's huge nose while remarking that Baker's "pet peeve is that everyone thinks he looks like the Sad Sack."

The opposite was true. Annie Davis, the pertly secretary who was the platonic love interest in the mainly male *YANK* office, went on dates with Baker on occasion, finding him shyly monosyllabic. Though his teeth were not scraggly, he resembled the ferocious sergeants and lieutenants who tortured the Sack. Living in the Bahamas in her 80s, she wondered why that was so. "Maybe some sort of mental quirk. I never asked."

YANK

The Girl 13 Next Door

On January 5, 1945, the final year of the war, Sergeant Baker's Sad Sack cartoon showed the tormented private under artillery fire, zigzagging back and forth between his own shelter and an evacuated German billet. Both in rubble, the two were alike except that a swastika adorned what was left of a German wall and a provocative female pose brightened the remains of the American dwelling.

Long since had been resolved the struggle between Mrs. Henry Stimson and her veto of a bathing beauty and the White committee's contrary counsel that *YANK* not be "too holy to engage in occasional leg art." Mrs. Stimson had lost, and American lovelies labeled "*YANK* pinup girl" were on display on barracks walls around the world. They were as familiar a part of the war experience as steel helmets and mess kits.

YANK had begun cautiously. For the first public pinup on June 17, 1942, the editors had selected a prim "girl next door," actress Jane Randolph who was described in the caption as a 125-pound lover of the outdoors and unmarried. In the next issue, the staff was a touch more daring. Flaunting a huge feather-topped hat, sleepy-eyed Ann Sheridan was presented as "Hollywood's number one oomph girl." Then, in issue number three, there must have been trouble. Jane Russell was the pinup. She was the star of Howard Hughes's *The Outlaw*, a movie so raunchy that the movie industry's self-policing censors forbade its release. Her sleeves rolled up but otherwise fully covered, the bosomy Miss Russell peered moodily at the viewer, and the caption urged every soldier not already equipped with a pinup on the underside of his foot locker lid to grab the chance to paste Miss Russell there.

Without explanation, the next 13 issues went without pinups. What behind-the-scenes battles went on were never disclosed, but on October 14, 1942, a pinup returned—amiably smiling performer Brenda Joyce wearing a one-piece swimsuit. *YANK* admitted that the actress's scanty attire was too late in the autumn season for much of the magazine's circulation, but it explained apologetically that where Miss Joyce had posed it was "warmer."

It was the last time *YANK* offered an excuse for a revealing shot. Evermore mindful of the White group's call for "occasional" leg displays, the weekly ran a total of 85 such studies, one in every second issue. One-piece bathing suits were the preferred attire, at first. Twenty such were published. Then the scantier two-piece costumes took over, 31 in all.

Except for two other occasions, each issue included a pinup. One such time was when Germany surrendered. For the June 1, 1945, issue, the mood was somber, the cover a blank page except for a three-inch-tall photo of a surrendering German, the faded enemy. The other time was September 7, 1945, with Japan vanquished. There was a female pinup in that issue, but it was one made of metal, New York's Statue of Liberty, the lady troops still stranded in Europe longed to see.

With the resumption of the pinups in the fall of 1942, Marguerite Chapman, of Columbia Pictures, was shown on November 4 in ladylike black gloves reaching to the elbows. Smoky-eyed actress Maria Montez appeared November 18 in an evening gown. But, not to worry. The magazine's temporary first editor smirked in a December 26 round-robin that "more sprightly cheesecake" was on the way.

A typical pinup girl from the magazine. (*Courtesy of* YANK)

Indeed it was. Hollywood studios, seeing the opportunity to provide patriotic home front support for the troops while advertising their wares, flooded *YANK* with offerings. Most were "sprightly." The one limitation was the Hays Office ban on bare bosoms, inner thighs, and lace lingerie. The studios complied with that, but the penultimate issue, in December of 1945, pressed the limits with Martha Holliday prone on her back, a strip of lace peeking from the panties. It was a final contemplation for the veterans to take with them as they went off to father the Baby Boom.

More than 100 actresses appeared in *YANK*'s pages. Eighteen studios and model agencies contributed. Columbia Pictures was the most prolific, with 30 winners, followed by Metro Goldwyn Mayer, Warner Brothers, Universal Pictures, Twentieth Century-Fox, and RKO, each with 20.

Many of the pinup posers, some long deceased, others in their 80s, are remembered now only by families and close friends, but others are still familiar thanks to movie reruns on television. Some posed early in their careers, taking on a different persona later as comediennes or dramatic actresses. Lucille Ball, famous subsequently on television, was the March 23, 1945, pinup, radiantly cheerful in an evening dress. Much-married Ava Gardner wore a two-piece bathing suit while tossing a beach ball on August 6, 1943. Lauren Bacall, then 20 years old, appeared in an evening dress on November 24, 1944, with a caption saying that her film work radiated "so much pure unadulterated sex" that some adjudged her the match, or the superior, of such earlier sirens as Marlene Dietrich, Greta Garbo, Veronica Lake, and Jean Harlow.

Ingrid Bergman, first a sainted screen portrayer of a nun, then involved in a scandal with her later husband, Italian director Roberto Rosselini, was a two-time pinup, with a smiling head shot on April 28, 1944, and a bare midriff pose beside a haystack on March 16, 1945. Singer Dinah Shore posed in an evening gown December 9, 1942. Dramatic actress Barbara Stanwyck showed another side of her talents February 19, 1943, in her scantily clad role as the "Lady of Burlesque" in a film inspired by stripper Gypsy Rose Lee's novel, *The G-String Murders*. Another dramatic actress pinup was Anne Baxter of Twentieth Century-Fox. On April 9, 1943, smiling gently, she posed covered top to toe.

Several others were repeaters. Jane Russell, of the suppressed

Outlaw film, who may have been the reason there were no pinups in 13 1942 issues, was back on January 6, 1943, clad in a low-cut dress and a smile. A half year later, no caption was necessary as the buxom beauty stretched out on a diving board, straining upward to grin into the camera. On September 21, 1945, by then 24 years old, in what appeared to be a negligee, Miss Russell became the pinups' only four-time winner.

Others seen more than once included the dancer Betty Grable, model Jinx Falkenberg, and actresses Marguerite Chapman and Gene Tierney.

In retrospect, the pinups were modest compared with the standards of later decades, but, with the war on, Mrs. Stimson was not the only one to raise eyebrows. At 95, Forsberg recalled that reports came to him, as *YANK*'s publisher, that in some latrines pinups had become agents of masturbation.

"Mail Call" comments about the pictures were enthusiastic, although occasionally critical. Among those hinting a shade of disapproval was a December 8, 1944, letter from Sergeant John Furwiller in Italy. He applauded the September 15, 1944, portrait of blond Conover model Betty Jane Graham, whose head and shoulders alone were shown. With a girl like that living next door, no one would move away, said *YANK*. Bravo, wrote the sergeant, don't stop the sexy pictures, but "give the rest of us a break" by also providing shots of "a typical American girl with the kind of beauty every man dreams about."

Private First Class Joseph H. Sailing, at the Myrtle Beach, South Carolina, Army Air Force base, said the same thing more bluntly. "I don't know who started this idea of pinups but they say it is supposed to help keep up the morale of servicemen. I would say that 24 out of 25 of the men in service are either married or have a girl at home who they respect and intend to marry." As for himself, the pfc said, he preferred for his wall a picture of a P-51 fighter alongside that of "the best looking girl in the world," his wife, rather than a view of "some dame who has been kidded into, or highly paid, for posing for these pictures." How would you like it if you were to find decorating the home of your wife or girlfriend a picture of "a guy stepping out of a bathtub draped only in a scanty towel."

Graceful Irene Manning, of Warner Brothers, became the catalyst for a showdown between the friends and foes of the pinup. A lovely person who would grace any country club, the fair-haired Miss Manning was portrayed seductively March 24, 1944, in a two-piece outfit as she lay on her side, a leg curled forward.

Sergeant E. W. O'Hara and Corporal P. Pistocco Jr. took exception from England. They could not see how "the intelligent looking Irene Manning" could have agreed to "such a suggestive-looking" presentation. "She is well dressed but that pose! Phew! Hays Office please take note."

For the phalanx of pinup lovers that was too much. Two four-stripe staff sergeants in Panama, Cliff Crouch and Raymond Cox, wrote in that members of their platoon thought O'Hara and Pistocco "should be examined for section 8," the clause in the code providing for discharge by reason of insanity. Pfc. Frank T. Lamparero, in the same theater, protested that studies like that of Miss Manning were "big for our morale."

"I'll take the pinups in seductive poses; can you blame me?" chimed in Corporal Anthony J. Barone, at Hunter Field, Georgia.

"If I had wife I would make sure her picture was up, but Irene Manning will do until that big day," offered Seaman First Class R. C. Walters, whose address was the Fleet Post Office.

Most heartfelt of all was a protest from two corporals, F. A. Wallbaum and Cooper Dunn, and three pfcs, Robert Ross, Lloyd W. Finley, and Elom Colden, in Alaska. They spoke from an experience parallel to that of the soldiers on Bougainville who were tongue-tied at the sight of nurses, the first women they had seen in many months. The five in Alaska expostulated, "maybe if those panty-waists had to be stuck out some place where there were no white women and few native women for a year and a half as we were they would appreciate even a picture of our gals back home. We nasty old engineers appreciate *YANK* with its pinups."

Prurient or not, *YANK*'s pinups were a response to the frustrations of thousands of men torn from a normal social life at the height of their youthful sex drive and, as for their daring, there was never one wearing a bikini, nor was there ever an unveiled umbilicus.

YANK

GI 14 Mind

In *YANK*'s first year, in the opinion of editor McCarthy, the third of a page devoted to letters from readers was "one of our lousiest departments."

By contrast, in the war's final year, there were more than two full pages a week devoted to "Mail Call." It had become even more popular than the pinups, rivaling in interest even the magazine's staff coverage of combat in Europe and the Pacific.

"Mail Call" provided an enduring record of what really went on in the minds and hearts of the soldiers, sailors, Marines, and Coast Guardsmen of "the greatest generation." Their reactions were closer to those of their grandchildren than the present generation may realize.

Like priming a pump, *YANK*, at first, had to write fake letters to itself in an effort to lure privates, corporals, and sergeants into telling the world what they were thinking.

In his May 11, 1943, round-robin, McCarthy recalled *YANK*'s early days when "everybody in New York wracked their brains" each week trying to fill the "Mail Call" space. In the first issue in June of 1942, for instance, "Private Homer Alexander," bearer of two great ancient Greek names, urged the magazine to hold down on corny jokes about potato peeling and "kitchen police" duty. *YANK* solemnly answered itself promising to seek humor elsewhere. Even months later there was something suspiciously provocative in a letter purportedly from New Guinea from Private A. Bocchino, translated "a little mouthful." Mr. Mouthful took note that women had joined the men in uniform as the members of the Women's Army Auxiliary Corps (the WAACs, later, simply the WACs, without the "auxiliary"). Nonchalantly content to outrage his sisters in the service, Mouthful wrote, "I don't like the way the (WAACs) look. They should be home knitting, that's my opinion. I thought this was a man's army but it seems I was mistaken. The only man's outfit left is the Marines. A woman has always been the ruin of a good man."

Tongue in cheek, pretending to take no part, *YANK*'s one-line answer to the chauvinist was that "there are gals now" also in the Marines.

Three weeks later Bocchino had an answer, either real or again staff written but, in either case, successful in drawing other readers into corresponding. "Corporal E. Haines of the First WAAC Training Center, Fort Des Moines, Iowa," responded with exemplary patience and dignity, "As for the way we look, we can't help that, God gave us only one face. We all realize we can never do the wonderful job our men are doing but we can try to help. The bigger part of us would like to be home if our men could be home too. As for your last uncomplimentary statement, a really good man can't be ruined by any woman."

Lest there be misunderstanding about where *YANK*'s own EM editors stood with regard to the female component, the magazine celebrated the first anniversary of the WAACs with a cover photo of six-stripe Master Sergeant Margaret M. Tootle, of the 36th WAAC Company at Fort Meade, Maryland. On the wall above her head was a sign: "This is my war, too."

Stimulated by such exchanges, *YANK*'s readers finally opened up, flooding the paper with opinions, fears, and appeals for help, an outpouring so candid and so critical of the officer corps that one serviceman at the Eagle Pass Army Air Force base in Texas informed

YANK that "this field has issued an order forbidding any of us from writing *YANK* of any of our gripes or complaints."

YANK insisted that every correspondent identify himself, but agreed to withhold the writer's name if exposure would embarrass him. Most "Mail Call" letters, however, appeared in print with the name and organization of the author.

A recurring theme was a protest against the Hollywood vision of life in the service. Staff Sergeant Michael J. Lawrence, at the Army Air Base in Herington, Kansas, raised the subject: "The tendency to glamorize the war is nauseating," he wrote on September 24, 1943. "And the film industry is perhaps the first offender. Men must die in war. Why can't the home front accept the fact without … covering it up with glamour."

Sergeant Kirk Faris, at Letterman General Hospital in California, agreed. "If you got your ideas from most movies and magazine articles," he said, "you would believe that every young man is just crazy about Army life." He denied that the typical GI volunteered immediately as soon as the Japanese attacked Pearl Harbor. He questioned whether stay-at-homes were "very bitter about not being drafted." He questioned whether those rejected for service cursed the doctors who turned them down. He scoffed at the idea that GIs "with a nice soft spot in a service outfit in the States are busting a gut trying to transfer to the infantry." He hooted at the thought that combat veterans who have been rotated home "having gone through hell, are dying to get back into a muddy foxhole."

"Don't get that last point wrong," the sergeant added, "it's commendable to be anxious for the fight. I'm only saying there aren't so doggoned many anxious guys as (writers) would have you think. Else why did they have to draft us?" As for those rejected as physically unfit for service, "I didn't see many" turned away at his induction center, and among those who were, "I haven't seen any who shed tears."

While the letter writers seemed to take for granted that a soldier had to do his duty, many challenged the idea that he enjoyed his miseries. When Private James S. Hamilton sounded a different note on December 3, 1943, he brought down a cloudburst of jeers. Hamilton asked *YANK* how he could be shipped abroad. He said, "I volunteered for the Army with the intention of fighting but I was put in the (military police) at Fort Custer, Michigan. I may have to guard defense plants. It looks as if I have a white collar job. I don't like it."

"Ha, ha, ha," replied Corporal Herbert S. Zinz and eight of his comrades in Italy. Private First Class Al Krupa and his fellow pfc Don North, in Algiers, used more words, "Listen, Hamilton, you don't know how well off you are. Any number of us over here would gladly swap with you. We eat C rations, sleep on rock piles, walk guard all night in mud up to and beyond the knees, and we haven't had fresh milk or ice cream since we left the U.S. Women over here aren't even white. They take a bath when it rains. You're sleeping in a barracks, talking to American women and enjoying a few hours off in the evening. Be satisfied."

What to do in combat was another topic of debate. *YANK's* Hawaii bureau chief, Sergeant Merle Miller, triggered one discussion with a piece, "Makin Taken," describing in matter-of-fact fashion what he saw when that islet of the Central Pacific Gilbert chain was wrested from the Japanese. Three of the enemy were shot down when a grenade flushed them from a shelter. Private Ralph H. Luckey, at Camp Davis, North Carolina, took exception, "As God is my witness, I am sorry to read of the way American soldiers treated the enemy on Makin Island. They shot some Japanese when they might have been able to take them alive. I know that if I were in a dugout and forced to run I would want mercy."

Letters of protest poured in, only two of them agreeing with the compassionate inductee. Veterans at the Norfolk Naval General Hospital in Portsmouth, Virginia, advised that "if Private Luckey heeds his own call for mercy, his soul will belong to God but his body will belong to the Japs." Sergeant Carl Bethea and 13 co-signers at a port of embarkation commented that "Private Luckey better live up to his last name if he goes into combat with the idea of taking Jap prisoners alive." Private C. E. Carter, at Harmon General Hospital in Longview, Texas, recalled lugging corpses out of the jungle during eight months on Guadalcanal and added, "Private Luckey will have no dead Japs on his conscience when they kill him."

The letters kept coming. "After being in combat and seeing medics killed trying to help our wounded makes you want to kill the bastards," Private P. Stupor offered to "that servant of God, Private Luckey."

"Please notify the FBI, G2 (Army Intelligence), anything, but have that guy locked up" was the comment of Corporal S. Schwartz of Fort Custer, Michigan. Even a namesake of the devout Camp Davis private wrote in to make clear that he thought differently. He was Second Lieutenant Ralph J. Luckey, in Italy, who was so anxious not to be confused with the other Luckey that he was willing to go overboard in the opposite direction. "The only good Jap," he said, "is a dead Jap."

Another furious subject of "Mail Call" discussion was whether troops had to salute German officer prisoners of war. It was kicked off March 10, 1944, by Sergeant Nat Reese, on duty at the Carson, Colorado, prisoner of war camp.

"Am I boiling?" he asked. "I looked forward to seeing the arrogant Nazis behind barbed wire but, damn it all, I've been ordered to salute all German officers." YANK checked with the P.O.W. section of the Provost Marshal General's Office in Washington and told Reese that, sad to say, the Geneva Convention of July 27, 1929, specified that "officers ... shall be treated with regard for their rank" and that, therefore, he had a new range of superiors to whom he owed a snappy salutation.

So many angry retorts came in that YANK went back for a second reading and, on August 4, 1944, reported that "camp commanders have now been notified that the Geneva Convention does not specifically require that captive officers be saluted and, therefore, that any local ground rules stating that enemy officer prisoners of war must be saluted is no longer in force." It was one of several occasions on which the EMs' publication seemed to have forced an Army change.

The outcries ran in several issues, but in one of them alone, May 5, 1944, there were 16. Sergeant William F. Wenzel, at Winter General Hospital, Topeka, Kansas, said that even at the risk of a court martial "I'll see myself in hell before I ... ever salute a German in a Nazi uniform while the United States is at war with Germany." An unidentified private at the Philips P.O.W camp in Kansas lamented, "We capture them and then we salute them—what a joke." Corporal Ernest T. Dutton, recuperating at the Kennedy General Hospital in Temple, Texas, said that, in eight months as a P.O.W. of the Wehrmacht in North Africa and Italy, he "never saw any Nazi guard or enlisted soldier salute our officers, British or American."

Another topic that sent excited soldiers to pen and paper was the issue of race relations in American society, and especially inside the armed forces. *YANK* itself raised the question in a letter December 17, 1943, which had all the earmarks of a staff-written provocation. Allegedly, it came from a Fort Ord, California, "Sergeant Nolan" and a "Corporal Hitner," the latter an evidently thinly disguised reference to Hitler and the Fuehrer's World War I rank. With redneck zeal, the two purported writers protested that blacks in the armed forces "think they are equal to the whites because they fought in this war." At the conflict's end, the two said, blacks "should not be given equality but should be rewarded perhaps with some portion of some country for their own use where they will be governed by white people in order that they will be taken care of and still have a chance to earn a good living."

The letter gave *YANK*'s editors an opening to protest in response that "it is folly" to say that the 153,900 blacks in the Army overseas, the 74,013 in the Navy, and another 488,961 in stateside Army units should be "governed by whites when (they) are fighting for our country's rights." It was a step toward proclaiming black civic equality, but what really made the matter a blazing issue was an apparently legitimate letter three months later from Sergeant James C. Hetrick, in Fort Jackson, California, and, more importantly, by a further letter, this time by a black GI on April 29, 1944. The letters and those they provoked showed how a new social consciousness was developing among those under arms, helping pave the way for the civil rights movement of later decades.

Hetrick, returning from a furlough, described what happened when his train entered a Jim Crow state. "As soon as we hit the state border all passengers were shuffled, whites in one car, Negroes in another. I saw Negroes told to stand up and move to another car, some of them veterans wearing starred (combat) ribbons. They were good enough to fight for our dear Southerners but they can't ride in the same car." It was Corporal Trimmingham's letter on April 29, 1944, however, that finally touched consciences. Writing from the hospital at Camp Huachuca, Arizona, he told *YANK* that "some of the boys are saying that you will not print this letter (but) I say you will." He was right. This was his story: En route to Arizona from Camp Claiborne, Louisiana, there was a one-night layover between trains in a community where "old man Jim Crow rules."

Trimmingham was part of an eight-man black soldier group. None of the town's lunchrooms would serve them so much as a cup of coffee but the railroad station restaurant agreed to give them a meal so long as they ate it in the kitchen. As Trimmingham looked on, two dozen German P.O.W.s and their two-man American guard were welcomed into the station dining room where they "sat at the tables, had their meals served, talked, smoked, in fact had quite a swell time."

The question in the mind of every black soldier "pushed around like cattle," said Trimmingham, is "what are (we) fighting for, on whose team are we playing?" Three months later, the black corporal filled *YANK* in on the reaction. He had been heartened by 287 letters "and, strange as it may seem, 183 are from white men and women in the armed service." He said he now had reason to hope for the best.

So many EM put faith in their soldier magazine that "Mail Call" spawned a twice-a-month, half-page feature inviting readers to confide "what's your problem." With the help of the Washington bureau and the Pentagon, *YANK* provided authoritative, though not always welcome, answers. Prisoners in lockups, soldiers with wife trouble, quarrelers seeking to settle bets—all turned to *YANK* as a super chaplaincy.

Many letters outlined worries about the family back home. Sergeant J. D. in Presque Isle, Maine, for instance, wanted to end his wife's allotment and get a divorce because she was "staying with another guy." *YANK's* advice to him was to check with one of the legal assistance officers, which had been provided by a War Department circular of March 16, 1943. Private Walter R., at Camp McCain, Missisipi, caught a venereal disease from his wife while on furlough. Would he lose pay during his time in the hospital? The reply was that he would probably not. Army Regulation 30-1440 suspended pay for time lost due to VD only when the GI's "own misconduct" was the cause.

Problems of every kind were reported. Her "loafer" husband wanted a $22 a month family allowance, said First Sergeant Mary Zither, in England. The answer to her was that her spouse could get away with no such claim at that moment, but that a bill then before Congress might yet meet his fond desires. Separated, but not divorced, Corporal O. T. R., in India, wondered whether his girlfriend could be in trouble for collecting an allowance as his wife. The answer was urgent, "Brother, you'd better sit down and write (her) a long letter. Come to think of it, mark that letter registered, air mail, special delivery." She needed to return everything and was at risk of a fine of $7,000 and three years in jail.

Those behind GI bars had their share of questions. From a private in New Caledonia came this: can you get veterans' benefits after a dishonorable discharge? The answer, no, not unless you wangle a special act from Congress and, if the DD was for desertion, nothing but a presidential pardon would even restore your citizenship. "Having been in the thick of the fighting, I find I want to be a chaplain," Sergeant Herbert Ranzall advised from New Guinea. If, as he understood it, the Navy allowed for theological training, would the Army send him to a seminary? Sorry, no.

Private George Benton, in Italy, wondered whether his "couple of months in the guardhouse" before shipping out of the States would be held against him when he qualified for discharge. Yes, was the disappointing answer. He would have to make up the missing time away from service before getting out.

Some letters revealed such weak personal backgrounds that they awakened memories of the exchanges between the New Yorker Hotel operator and the Texan in medics' basic training in Camp Barkeley. "My buddy and I," Private Edward W. Schultz wrote from Italy, "had an argument about the century we are living in today. I claim we are living in the twentieth century and my buddy claims we are living in the nineteenth century. Now tell me who collects the ten bucks we bet."

Said *YANK*, "You win but you should be ashamed to take the money."

Our Allied sisters in arms seemed no better informed. Private Asa W. Willis, in the British forces, asked help with "an argument with several of my friends in camp." The question was how many states make up the United States of America. Thirteen, said one. "I said she was thinking of the 13 colonies when America was first founded and I told her that there were 48 states. Hoping I'm right. Cheerio."

Ever ready to help readers wherever they were, *YANK* settled the

English ladies' debate, "We started out with 13, but we've gradually accumulated 35 more."

After its rocky start, *YANK* clearly had won the confidence of the enlisted soldier.

YANK

Gallows 15 Humor

Just as Eastern Europeans under occupation by the Soviets during the Cold War cheered themselves with "gallows humor," the Poles saying when the Russians launched a dog into the sky at the beginning of the space era, "We, too, are half way there, we have dogs," so did the half dozen cartoons in each *YANK* issue offer escape to anguished soldiers. Often poking fun at the officer corps, cartoons frequently were comic metaphors for real sources of GI grief.

The high-spirited staff cartoonists and artists, many of them art school graduates and professional illustrators, had their own partitioned-off corner in the *YANK* city room, a space the now 87-year-old ex-sergeant Greenhalgh remembers as a lighthearted, irreverent sanctuary rarely invaded by the first sergeant or the officers. It was noisy.

"Artists are noisier than writers or accountants," Greenhalgh recalls. "It must have something to do with the manual nature of the work. Artists can talk and draw at the same time. I never saw a writer who could talk and type."

Occasionally, the artist pranksters ventured out of their quarters to cause small havoc in the city room. Annie Davis Weeks remembers how the sketchers one day lighted a fiery powder train around the chair of the startled McCarthy.

Although the *YANK* cartoonists ruled out KP jokes as hackneyed,

135

they found plenty of macabre laughs to replace them: parachutes that failed to open, shipwrecked sailors on desert islands, submariners in trouble, unlovely half-clad wenches in the tropics, and risky experiences aboard carriers that had no railings. As more and more of the artists shipped abroad, the officer corps became a more frequent target.

Some who have pondered the lessons of World War II, such as movie director David Brown, conclude half in earnest that, the Axis to the side, "the real war was between the officers and the enlisted men." The *YANK* cartoons suggested that. With a relative handful of professional service personnel, the graduates of the Annapolis and West Point academies on the one side, and more than 10 million newly inducted civilians on the other, friction between the two was inevitable.

This second phase of cartooning began with soldier frustrations. Sergeant Jack Lovell, on September 17, 1943, portrayed a naked soldier saluting an astonished officer. He had been told to come at once. "Brown, sir," he said, "reporting as ordered—without delay."

Supplementing the graduates of the service schools and of National Guard units were the products of the colleges' ROTC and the Army's OCS, many of them just out of their teens and in command of EM former executives in their 40s. Grinding the teeth of many inductees were shavetail (second) lieutenants and other young officers, proud of the gold and silver bars on their shoulders. A rash of cartoons was the result.

The first appeared in the August 12, 1942, issue, the work of Private Douglas Borgstedt, a 31-year-old former cartoon editor for the *Saturday Evening Post*. He was a part of the original *YANK* crew that put together the inaugural issue. He sketched a husband and wife in bed, with the woman protesting to her captain spouse, "Harvey, you are not going to wear those bars to bed!"

A flood of others on the same theme of young bosses and

mature EM followed. Sergeant Earl Carver, in Panama, limned a matron at a reception asking amiably, "What are you going to be when you grow up, lieutenant?" Corporal Ernest Maxwell on September 1, 1944, sketched a stiff-backed young ensign counseling an old sea dog at a gangplank: "For Mother's sake, Dad, try to behave ashore." On December 7, 1945, it was Maxwell again, this time with a corporal asking a lieutenant, "What shall I do with your comic books, sir?" Still more, on January 5, 1945, Sergeant John W. Frost had a grimy technical sergeant of the Air Force ground crew in an embarrassed heart-to-heart talk with a baby-faced captain: "Sir, I guess the best way to explain it is to start with the bees and flowers."

All in the services, from sergeants and applicants, to OCS, on up to colonels and multi-starred generals, became targets for the pictorial scoffers, with the sketchers sometimes skimming the edge of sedition. Eventually promoted to five-stripe technical sergeant, and content to remain as an enlisted man on *YANK*, Borgstedt took a poke at fellow EM seeking to cross the divide into the officer class. What type of people were they? He drew a jungle scene with seven candidates lined up to apply for OCS. Fourth in the list was a vacant-faced gorilla.

In 2000, at 89 years of age, shortly before his death, Borgstedt spoke of that sketch and the freedom *YANK*'s cartoonists enjoyed:

YANK gave all GIs a big boost in the never-ending relationship between enlisted men and the officer corps. It made most GIs feel "we're just as good as you are," and it put a big dent in the invisible wall between GIs and officers. ... *YANK* pretty much did what it wanted to, witness my gorilla cartoon.

As weeks went by, the humor in the cartoons began touching on serious grievances about command insensitivity, the misuse of power, and the abuse of underlings. On August 4, 1943, Sergeant Irwin Kaplan, at Fort Knox, Kentucky, showed a happily self-satisfied brigadier general greeting a bedraggled soldier who had just brought in 100 German prisoners. "Great going, Mahoney," the conqueror was told. "You'll get pfc, for this." In a latrine in the November 16, 1945, issue, Sergeant Ted Miller had a full eagle colonel rewarding a hard-scrubbing Private Smith for excellent efforts, "I'm having you transferred to the officers' side!"

YANK's first cartoon shot at the officer corps was aimed safely overseas with an enemy as the target. Captioned "Giuseppe in trouble," it was the work of Borgstedt and was published August 5, 1942. It portrayed an Italian private, Giuseppe, sprinting away from combat and just managing to outrun one of his own generals.

Two weeks later, Borgstedt aimed his mockery at an officer of less than general rank, this one an American. In rough terrain the officer was bumped high out of his jeep, but not so loftily as his driver, a private. "Dammit!" he protested. "As your superior officer, I insist on bouncing higher than you do."

Sergeant Dave Breger, in *YANK*'s initial issue, launched a comic strip, "GI Joe," which joined Borgstedt in making fun of the brass. In the September 16, 1942, issue, Breger had a lieutenant admonishing a private who was filling a coal bin to do so quietly, one lump at a time, since "the colonel's about to take a nap."

Breger's anti-officer drumbeat took on a steady rhythm. On September 23, a dim-witted officer in foppish riding breeches warned GI Joe to take better care of the jeep tires. "Do you think rubber grows on trees?" he wanted to know. In the following issue, a major told his dubious troops not to be yes-men, "even if it means court martial."

With America's own generals treated gingerly at the start, *YANK*'s EM gathered courage and soon were giving the starred officers the full blast of their corrosive attention, something no other army could imagine. There was every type of put-down.

Sometimes the EM's own worries were turned on their heads with the generals paying the price. In every unit there were privates

YANK cartoons by Sgt. Douglas Borgstedt. (Courtesy of YANK)

unable to become pfcs and corporals who could not make it to sergeant because of the T/O (table of organization). However frustrating, the T/O had its logic, in any tribe there can be just so many chiefs and, necessarily, far more Indians. It gave Pfc. Irwin Touster the idea for an April 14, 1944, cartoon: a two-star general comforting a miserably gloomy one-star brigadier, "Maybe we can get something for you when we get our new T/O."

139

Along with promotions, there was the question of decorations, "fruit salad," as the multicolored chest ribbons were called. Who gets them and how fairly were they distributed? Touster, by then a corporal, had his doubts about the equity. His December 21, 1945, cartoon, at the eve of *YANK*'s dissolution, had a walleyed major general decorating a peer with the remark, "Don't forget, Pulling, you write one up for me tomorrow."

Injustice with regard to whom should pay when things went wrong was the topic for Sergeant Miller on November 16, 1945. A one-star general chided a full colonel for a "very poor inspection" of his quarters, "shoes dusty, bed messy." Result? "I had to restrict your orderly three days."

Misdemeanors could cost an EM a pass away from camp, and could get him restrained to quarters, so on September 10, 1943, Private Art Kraft tried that one at the generals' level. Two double-starred generals conversed, one of them in bed. "I've been restricted to quarters again," the recliner said. With medals on his pajamas and federal eagles atop the posts of his canopied couch, the brigadier had a food tray on his lap.

Rank has its privileges and those enjoyed by the brass may have been much less lavish than envious EM thought, but that did not keep Sergeant Jim Weeks, on May 18, 1945, from portraying a high-living brigadier general suffering from gout, his foot in bandages. "Trench foot?" asked a visitor, giving the over-luxuried warrior the benefit of the doubt.

In an Army famous for the need at most times to "hurry up and wait," a source of discontent was the many days waiting for shipping, both during the time of the fighting and then, after the Japanese surrender, trying to get home. Corporal Touster on November 30, 1945, got an idea from that. "It's this uncertainty and sweating out transportation that gets me down," one major general told another as the two enjoyed a Scotch and soda tipple.

Although *YANK* saw to it that none of its staffers stayed forever on 42nd Street, many thousands in the armed forces passed years in rear echelon jobs far from the sounds of combat. High brass were among them, happily distant from harm's way. On August 10, 1945, in the interlude between the two surrenders, Sergeant Weeks, with a touch of bitterness, worked that theme: two obese officers in conversation evidently in Washington, a bucktoothed brigadier general inviting an eagle colonel to feel the weight of his paper-stuffed carrying case. "You really know there's a war on, just feel that briefcase," was the caption.

An irony in the service was the inflated importance of the old Regular Army troops, including Dwight Eisenhower himself, obscure low-ranking personnel in peacetime suddenly in the crisis of war catapulted into positions of vastly inflated importance. Several cartoonists had fun with that. On April 23, 1943, Sergeant Kaplan depicted a slim, shaven-headed young lieutenant awkwardly asking a full-bellied, multi-ribboned, four-star full general, clearly a Regular Army man, "What did you do for a living in civilian life before the war, general?" Sergeant Tom Flannery on December 7, 1945, had a dull-faced six-stripe military old-timer asking a pfc, "This civilian life everyone's talking about, what was it?"

For some of the war's greats, including many who came from the Regular Army and would return to it, there was no further promise of glory, as Private Willard G. Levitas noted on September 29, 1944. Two officers sipped the omnipresent drinks, something combat EM could not have. "Objectively speaking," said a double-barred captain, "the postwar period holds no terrors for me. I can always go back to my permanent rank of pfc." Sergeant F. Phillips, on April 27, 1945, used the same theme from a different perspective. In the Western Union messenger force, the boss was shown objecting to the wishes of one of his carriers, a former second lieutenant, "None of the others insist on wearing their bars."

As the war ended, some of the cartoonists were still bitter, even menacing. In the last issue on December 28, 1945, Touster, still a corporal, had an eagle colonel tasting wine in a shop. "Excellent," he told a captain. "Put this place off limits at once."

In the issue just before that, Pfc. McNaughton, at Camp Cooke, California, showed a thug, evidently an ex-GI, staring at a captain and asking, "Haven't I seen you some place before?"

Even more sinister was Pfc. Walter's offering on September 28, 1945, a fellow one-striper leering at the remains of an eagle colonel and saying, "Now we can be friends." In the colonel's head a hatchet was imbedded.

YANK was an official War Department publication, a part of the Army. Authorized by *YANK*'s EM editors the cartoons, irreverent, even outrageous, were published. In an authority-focused organization they were a contradiction in terms, but in such a civilian-manned army they served a cathartic purpose. Against military logic, they helped win the war.

In an implicit tribute to the wit and wisdom of so many of the cartoons, *The New Yorker* magazine, itself famous for its humorous sketches, asked to see and use the castoffs *YANK* did not publish. *YANK* obliged.

YANK

Unpolished 16 Brass

As *YANK*'s cartoonists played all of Army life for a laugh, making it clear that they had no intention of polishing the brass, some serious questions about EM–officer relationships confronted the enlisted editors.

What should *YANK* do when some of its talented correspondents were awarded with battlefield commissions? Could such staffers stay on the staff?

Allowing the "Mail Call" columns to air GI grievances against superiors might be an end in itself, releasing some hot air and, at least, assuring everyone that he had a friend who would listen. But what if there were grave allegations? What in that case was the magazine's duty?

Finally, should *YANK*'s devotion to the EM be so blind that great stories about officer heroism be ignored?

Before the end of *YANK*'s first year, Staff Sergeant Robert Neville, in North Africa, and Sergeant E. J. Kahn Jr., in New Guinea, presented *YANK* with the first of three problems.

Bob Neville had been the bridge columnist for the *New York Herald Tribune*. At ease in every social situation, he had never adjusted to the status of a GI. He refused to wear his telltale stripes and even treated generals in a familiar, if not condescending, manner. In London, he interviewed Lieutenant General Andrew "Andy" G.L. McNaughton for the December 30, 1942, issue, and used part of his full-page spread to commend the three-starred officer for the

143

precision with which he used language.

"I was impressed," the sergeant shared with his GI readers, "with the consummate accuracy of (the general's) talk."

General McNaughton was commander of the Canadian troops in Britain. Bob called the piece "The General Called Andy."

Although generals did not customarily dine with staff sergeants, Bob found it appropriate to explain to his *YANK* audience why he and "Andy" did not break bread together after their one-hour morning conversation. The officer's aide, presumably apologetically, said, "The general won't be able to lunch with you." Andy had just been notified that one of his three sons had died in a raid over Germany.

In between sending *YANK* a steady stream of dispatches from Australia and New Guinea, "all of them printable," according to the first editor, Major Spence, Jack Kahn had just found time to publish a book, *The Army Life.* It was "selling well" with critics agreeing that it "approaches literature," Spence was happy to report in a late 1943 round-robin. Both Kahn and Neville were praised in another letter to the field for fine "battlers," good combat reports, Kahn's from near the equator, Neville's from Oran in Algiers. Then came the crisis. Neville and Kahn both crossed the line into the ranks of the officers while others soon followed. Paris-based Sergeant Robert Moora, a former *Herald Tribune* colleague of Neville, became a lieutenant, Navy Yeoman Third Class Allen Churchill an ensign, and even the officer-bating cartoonist Breger jumped up to shavetail.

At Fort Bartholomew, there was soul-searching followed, finally, on March 12, 1943, by an editorial that set the course for the rest of *YANK*'s life. It read:

> A couple of *YANK* staff members overseas got themselves promoted to officers. ... We did a lot of arguing around the shop one way and another because these men ... are charter members of the staff and excellent reporters. In their defense some of our staff pointed out that *YANK* had printed

poems by officers, cartoons from officers and, on one occasion, a feature story by an officer. ... But the idea of an officer as a reporter kills the very heart of *YANK*. ... It is therefore with regret ... that we say goodbye and good luck to Lieutenant Neville ... and to Warrant Officer Jack Kahn. Neither officer will henceforth do any writing for *YANK*. And just to put the lid down tight, *YANK* in the future will not print any contributions from officers, not even poetry. This is an enlisted man's paper. We are going to keep it that way.

Neville became the commanding officer of the *Stars and Stripes* newspaper in North Africa and in Italy, and rose to colonel. With no hard feelings, he allowed his staffers to write for *YANK* when the magazine was shorthanded. He was also generous in offering commissions to others. One who turned him down was Master Sergeant Herb Lyons, the executive editor of the Mediterranean edition of *Stars and Stripes* and, in the opinion of *YANK*'s editor McCarthy, one of the top handful of Army journalists. McCarthy, in his August 5, 1944, round-robin spoke of a conversation he had with Lyons. The two saw eye-to-eye:

It makes a difference when your staff has no officers. That is the only way to run a GI paper. You don't need a commission. If you want something done that requires rank you can always get an officer to do it for you. It is very important that correspondents of the *Stars and Stripes* and of *YANK* not have the officer privileges the civilian correspondents get. They should have the same news coverage opportunities but they should stay out of the officers' mess and live in the field. It is the only way to keep the GI point of view. You cannot reflect the views of the enlisted man unless you eat the same food he does and take the same hardships he does.

All the newly minted officers were expelled from *YANK*. Breger's "GI Joe" vanished from *YANK*, but First Lieutenant Breger continued as a cartoonist, this time for *Stars and Stripes* in London, a largely GI newspaper with no qualms about an occasional officer as a contributor. Neville stayed on in the Mediterranean and the Middle East after the war as a bureau chief for *Time*. He spoke nostalgically of going back to the States, but Rome was his final residence.

The matter of what *YANK* should do when "Mail Call" received serious accusations against one of the brass came up on May 1, 1944, in a letter from Private Joseph Peter Pinkos of Camp Dorn, Mississippi. He wrote:

> I challenge you to print this letter. ... I won't go very much into details; just a little general sketch will be enough. ... First, does anybody in the United States Army have the authority to whip a man with a hose? This is what is happening here in the Camp Dorn stockade. Yesterday nine men were whipped for minor offenses. One was whipped so badly he went to the hospital in an ambulance. ... Is this the kind of treatment they dish out in all stockades or is this an exception? I myself got 24 lashes; for what, I don't know. Why do things like this exist in America?

Though he denied it, one could presume that the private had some inkling of why he fell from grace, but *YANK* did not quibble about that. The letter was postmarked April 30. Seventy-two hours after Pinkos dropped it into the outgoing mail, *YANK* saw to it that it was on the appropriate War Department desk. For more than two months, the "Mail Call" editor delayed publication. On July 7, *YANK* received an official comment, "The War Department says that as soon as complaints of these conditions reached responsible authorities, disciplinary action was taken. Court martials have resulted." Fourth Service Command headquarters in Atlanta was a shade more forthcoming. The military courtroom proceedings, it said, produced evidence that there had been a "misassignment," as well as "incompetency." News reporters found out still more, and, on August 4, *YANK* ran the whole story, including what the civilian journalists had learned:

Major Louis Rothschild Lefkoff (was) court-martialed on a charge of standing armed guard ... while MPs under his orders flogged the bare backs of nine GI prisoners with rubber tubes weighted by .45 caliber bullets. Major Lefkoff was found guilty of the charges and sentenced to forfeiture of pay and allowances, dismissal from the Army and confinement for one year at hard labor.

YANK was not a subdivision of the MPs. Its job was to root for the EM, not to police the officer corps, but the Camp Dorn case was an exception.

What about officer heroism providing great adventure stories? McCarthy addressed this subject in his staff letter on May 3, 1943. "Some of you," he wrote, "probably have an understandable tendency to turn away from fighter plane stories because the men involved are officers. However, it is well to remember that (they) are excellent copy and we can use such stuff now and then even though it concerns lieutenants."

What brought it up was "a peach of a story" 42nd Street had just received from Assam, India, from two sergeant staffers, Cunningham, formerly of the Infantry, and photographer Bob Ghio, an ex-MP.

"This is an enlisted man's paper and all that," said the level-headed McCarthy, "but if you ever get hold of a thriller like (the Cunningham-Ghio piece) don't turn it down because the characters happen to be cursed with commissions."

Entitled "The Assam Dragons," the story ran as a three-page lead on May 28, 1943. It told what happened when Lieutenant Melville Kimball's plane conked out in Burma during a flight from China to India. Kimball thought he had made it safely back inside Allied lines in the subcontinent, but, in fact, he had ditched "in the middle of a well-populated Japanese advanced base."

Overhead, his fellow "Dragon," Captain Charles H. Colwell, observed Kimball's plight. Colwell's own plane was in poor shape.

Five bullets had cut through it. Nevertheless, he put off his own return to base and began strafing the perimeter of the downed pilot's position. That kept Japanese from reaching Kimball, but Colwell's gas supply was running low. He radioed to Assam, and other members of the squadron arrived to relieve him. With the strafing going on, First Lieutenant Ira M. Sussky offered to fly a PT-17 trainer to Kimball's clearing to rescue him. With enemy bullets peppering the air around them, the two pilots, still protected by a wall of strafing, chopped down brush to lengthen their runway and then, after eight unsuccessful tries, succeeded in getting the trainer into the air. All were saved. In a last order of business, the Dragons shot up Kimball's P-40, keeping it from the enemy.

YANK

That **17** Crazy George

Though *YANK*'s EM enjoyed themselves by generally ignoring officers in their copy and jeering at generals in their cartoons, in reality it was the officers with shoulder stars who held much of *YANK*'s fate in their hands. In mid-war, three of them, with a dozen stars among them, had a falling out concerning what should be done about the by-then-popular publication.

They were Lieutenant General George "Blood and Guts" Patton Jr., commander of the Seventh Army in the African/ European theater, and General Douglas MacArthur in the Pacific, neither of them happy with Sergeant McCarthy's product. Those two were on one side, and General George Catlett Marshall, the Army's chief of staff and overall commander, was on the other.

Marshall made a practice of reading each issue of the magazine, and generally liked what he saw. He was in full agreement with repeated orders from the secretary of war that the civilians in uniform, the millions of enlisted men, should have their own global periodical in which they could share views about the service as they experienced it.

Patton and MacArthur, on the contrary, saw their role as uncomplicated as the defeat of the enemy, doing that with a disciplined organization run from the top. As referee, Marshall succeeded in defending *YANK* against Patton, but managed little more than a draw with the "Commander Down Under."

In North Africa, Patton protested to Washington, even about some of *YANK*'s coverage of the war half a world away in the South Pacific. It was doctrine with Patton that neatness and gallantry go hand-in-hand; an unkempt soldier equates to a bad fighter. It was reported that Patton even insisted that the crews inside his tanks wear neckties. With that in mind, Patton was said to be horrified to see pictures in *YANK* of sweltering South Pacific troops growing beards.

What outraged him more was the cover image on the April 9, 1943, issue, showing Corporal Marcus E. Barrett, of Nitta Yuma, Mississippi, relaxing over lunch at an advanced base in the North African desert. If cleanliness was the ticket for military advancement, he was one corporal who would never don a third stripe. Barrett's elongated GI cap was pulled down round as a bowl, and the sharp eye of the camera had caught at least a hundred stains and dirt spots on his baggy pants, his shirt, and his shoes. He was a poster boy for sloppiest man of the year. With a face dark with fatigue, he hunched over a packing case as his table.

YANK's five-stripe Sergeant George "Slim" Aarons had snapped the picture not to glorify the unsanitary, but to salute devotion to duty. Barrett was the unit cook. He had fed the troops and, as the caption pointed out, was now "the last to eat." Dedicated to duty or not, Barrett was still a spectacular sight.

McCarthy recalled later what followed. "He (Patton) went berserk and sent a fiery cablegram immediately to the War Department, complaining that *YANK* was encouraging dirt and disorder among the troops and needed strict censorship."

YANK had to move fast to protect itself. Colonel Forsberg went to work. A peacetime publisher who believed in candor in periodicals, he had also come to understand that only the brass could deal effectively with brass inside the Army, so he had already worked out a system for such emergencies. Forsberg had gained the friendship

Aarons's photo of a sloppy cook, which outraged General Patton. (*Courtesy of* YANK)

of a sympathetic writer who was a brawny West Pointer as well as the aide of morale chief General Frederick Osborn. Forsberg's special academy graduate was Colonel (later General) Lyman Munson, godfather of one of the two Forsberg sons. West Pointers, the *YANK* commander philosophically had accepted, were the only ones who could reason successfully with other graduates of the Hudson River military school, so Munson was of crucial importance.

Osborn, for his part, had been a Wall Streeter in private life; a colleague of another financial district operative, Henry Stimson; then the secretary of war and immediate superior of Chief of Staff Marshall.

Forsberg sent his defense of Aarons's photograph, editor McCarthy, and the rest of *YANK* up through the circuitous route of friendships to Osborn, and that worried general—a star short of Patton's three—carried the censorship demand to the chief of staff. It turned out that there was no need for alarm. As *YANK*'s EM heard it, Marshall told Osborn, "Don't pay any attention to that crazy George Patton. Tell those boys on *YANK* to keep putting out the magazine just the way they've been doing it."

YANK had a few more brushes with their ETO critic, but they were all at the magazine's own initiative. When Patton slapped a sick soldier in North Africa who was causing widespread protest, Colonel White, *YANK*'s original commander and by then chief of the *Stars and Stripes* daily in Algiers, ordered his staff to omit the story. Scathing publicity in the civilian media already had been "punishment enough," White explained to war correspondents. Even that remark, however, gave new life to the controversy that brought *YANK* into it. "It kept our telephone ringing," McCarthy wrote to his scattered staff. "The United Press and AP and all the papers here started calling up to find out if *YANK* was going to soft pedal the story too. We assured them that we were going to run it."

YANK did so, actually defending the Army commander, but doing it in such a roundabout way, and with such a questionable instrument, that it was almost as embarrassing as it was helpful. A letter was run in "Mail Call" from "Private A," a prisoner in the Jackson Barracks, Louisiana, guardhouse, who identified himself proudly as "a soldier with nine years' experience." He wrote, "I heard General Patton was on the carpet in front of Congress. Why the hell don't they mind their own business and leave Patton and

Eisenhower alone? When a man is frightened there are several ways to bring him around and this could have been Patton's idea."

If that was aid for the magazine's nemesis, an editorial on October 20, 1944, was not. It was entitled, "A $1,000 Bill." It began:

> There has been a lot of space in the newspapers lately about the bet that Lieutenant General ... Patton is supposed to have made when he got to France. The general allegedly hit the Normandy beaches waving a $1,000 bill and betting that he would get to Paris before generals (Field Marshal Sir Bernard L.) Montgomery (of the 21st Army Group) and (Lieutenant General Omar N.) Bradley (of the 12th Army Group).

Patton, the article admitted, has "completely denied the story, adding that he has never seen a $1,000 bill," but if that ended the matter for some people, it was not so for the anonymous editorial writer. A group of "Texas citizens," he reported, had taken up a collection "so that the general will have a $1,000 note to wave when he rides into Berlin."

It was "bad taste," the editorialist protested. "This war isn't a game." Four hundred thousand Americans already were missing, wounded, or dead. "This business is more than a $1,000 bill waved at Berlin."

As far as bad taste was concerned, there was the question of whether or not it made sense to imply General Patton's guilt in the face of his denial, and how inappropriate it was for an EM safe and sound on Manhattan's 42nd Street to speak of the dangers Patton and his forces faced on the very eve of the "business" of the Battle of the Bulge. It left the suggestion that in the *YANK*-Patton struggle, the magazine's EM were not necessarily always right.

So far as the editorials were concerned, EM explaining warfare to generals, the War Department, with its hands-off policy, allowed such commentaries to be published, but the master edition thought so little of them that it never asked any overseas edition to run them.

The intermittent feud with Patton persisted to the end. On September 28, 1945, with both wars won, a cartoon by Sergeant Michael Ponce de Leon, of Scott Field, Illinois, was published portraying a bedraggled, slack-shouldered pfc about whom two of his fellows remarked, "His name is Patton. We call him Blood and Guts."

A problem with General MacArthur was that he wanted no civilian interference in his area and, as quickly became apparent to Forsberg, the general saw the New York master edition of *YANK* as all too civilian.

There was another problem, as the alumni magazine of *YANK*'s sister publication, *Stars and Stripes*, said in February 1994 about their own Manila edition in the MacArthur area. That was the fact that even if such a publication as *YANK* was an official War Department periodical to be used in all theaters, the strategically brilliant but hubristic general in Sydney, Australia, did not see himself bound by that since "he was subordinate only to God."

An early hint that *YANK* would have trouble in "MacArthurland" came when three *YANK* sergeants—Dave Richardson, Ladd Haystead, and John D. Harrison—presented their orders to the general in late 1942. Signed by General Marshall, they gave the *YANK* EM permission to travel as they pleased while living away from barracks on the civilian economy. Richardson remembers the general's reaction. "Well," he commented wryly, "all I can say is that you will have more freedom than anyone else in my command except me."

That swallowing of such an arrangement would be difficult for the general soon became even more apparent. *YANK* was allowed to set up a local edition, but not under the name used in other theaters; *YANK, The Army Weekly* became *YANK Down Under* (*YDU*). Handling editorial chores, Staff Sergeant Harrison told the New York office that Major Harry E. Berk, speaking with the general's authority, had told him to round up whatever local GI editorial talent he needed to put out "a strictly Australian edition." When a staffer wanted to cover combat for the world *YANK*, he was told by the local brass to busy himself instead with rear echelon copy of *Down Under* interest.

A problem was that the shorthanded MacArthur had thus far no publication of the *Stars and Stripes* kind. While *Stripes*, too, was staffed largely by EM it was in each area a local newspaper dedicated to the interests of the particular command. It was not a worldwide forum for EM talking to distant fellow EM. After Sergeant Harrison's worried report, it was evident that the MacArthur area wanted *YANK Down Under* to be, at best, a cross between the world *YANK* and *Stripes*.

The *YANK* rule was that each local edition could scrap six of the 24 pages of the master edition, dropping in material of local interest. The staff of the Australian *YANK* was pressured to drop an extra two of the New York pages, and, when Sergeant McCarthy analyzed what he said was a typical MacArthur-area issue—that of August 27, 1943—he found that eight and a half pages of the master edition had been scrapped. Both the cover photo and the lead article had been replaced. In the view of *YANK*'s world headquarters, the inserts not only were too narrowly local, but they were also below the magazine's professional standards.

With few of *YANK*'s EM aware of what was going on, a struggle over *YANK* in the MacArthur area went on at the level of colonels and multi-starred generals through the second half of 1944 and, even to a small degree, to as late as VJ Day, the time of the Japanese surrender. At that time, Forsberg was still on one of his trips to the Far East, trying to settle squabbles.

A major months-long scrap began May 5, 1944, when the staffs of Osborn and Marshall arranged for a worldwide instruction with regard to the content of the various *YANK* editions; it read: "In keeping with *YANK*'s global nature, the War Department considers it important that material published in the foreign editions of *YANK* be concerned primarily with the worldwide activities of the armed forces." No edition should be focused mainly on one area.

"Not so," was the immediate retort from the far Pacific. Effective this day, May 14, 1944, MacArthur's deputy chief of staff, two-star general Charles P. Stivers, told Osborn as *YANK*'s overseer, his theater commander wanted it accepted that *YANK Down Under* was "a publication of the United States Army in the Southwest Pacific Area" and no longer a subsidiary of the War Department's *YANK*. MacArthur would control all aspects, including top personnel, editorial content, and finances. New York's *YANK* would be free to offer features and other materials for *Down Under* consideration. In return, *YANK Down Under* would supply theater reports to New York.

Impossible, was the response from *YANK* in New York, seconded by the Osborn and Marshall staffs. The home office logic was spelled out on May 28, 1944, in an analysis of *YANK*'s envisioned role written by Major Jack W. Weeks and sent up through channels to the Marshall staff. Weeks had been a prewar organizer in Detroit for the American Newspaper Guild, the reporters' trade union. In mid-war, he was *YANK*'s executive officer, and at war's end, with Forsberg back in civilian life, *YANK*'s final officer in charge. Weeks pleaded the case for one worldwide *YANK*:

> *YANK* was founded with the idea that the fundamental interests of enlisted men are the same throughout the Army and that the enlisted men of this Army should have an official publication of their own. ... (It was) conceived as a world-wide operation which has, or should have, a consistent approach throughout all of its sub-operations in the various overseas commands. ... It is believed necessary that the men of the Army no matter where they are stationed should have a universal point of view and should know about and appreciate what their fellow soldiers are doing in areas other than their own. ... It is believed to be most undesirable for men to return from an overseas theater with all the prejudices that frequently accompany isolation, believing that they ... alone made the sacrifices and did the important job. No matter

156

where they serve, men are inclined to become provincial and to develop a state of mind contrary to the best interest of the Army and the country as a whole.

That said, Weeks turned specifically to *YDU* and how, under MacArthur's demands, that edition was "notably failing to fulfill" the publication's mission:

> Provincialism is being encouraged in that area by an overplay of news about the Southwest Pacific Area. The cover of *YANK Down Under* and the lead articles invariably pertain strictly to SWPA regardless of whether during that particular week the activities in SWPA were as important as the activities in the other areas. This is not only very poor news practice but it fosters a lack of respect for the efforts of the American soldiers in other equally important areas and could contribute to serious postwar friction.

Weeks may have been too pessimistic about the inability of the *YDU* readers to get along after the war with the veterans of the ETO, but he was more prescient than he knew about how geographical isolation could impact negatively on MacArthur himself. Despite magnificent successes in Australia, New Guinea, the Philippines, occupied Japan and Korea, and the rise to the presidency of his ETO peer, Dwight Eisenhower, MacArthur's brilliant career and his struggles with Washington were to end with his inglorious sacking by Harry Truman.

With neither Marshall nor MacArthur succeeding completely in getting their way, the staff of *YDU* struggled along, trying to serve two masters. The irreverent general-baiting cartoonist Borgstedt served three years as art director for *YDU* and, at age 90, just before his death, reminisced about it:

> We usually got the same privileges as civilian war correspondents—as we were entitled to by War Department orders—in attending news conferences and

covering events, but there were some big exceptions. General MacArthur refused to be interviewed by a *YANK* correspondent feeling it was below his status and dignity to go one-on-one with a lowly enlisted man. This really pissed us off as Admirals (Chester) Nimitz and (William "Bull") Halsey and top Marine generals cooperated with *YANK*.

The battle over the enlisted men's publication under MacArthur was a prime headache for the magazine, but at least it was an exception. On July 4, 1944, as Osborne wrote tartly to MacArthur's Stivers, one two-star general to another, "We are printing *YANK* in twelve overseas theaters altogether and in each the operation is successful and, with the sole exception of the SWPA, mutually satisfactory."

The out-of-country editions were desirable as a way to increase circulation, but, more than that, they were crucial to the accomplishment of *YANK*'s assigned mission to bond together around the planet what Sergeant Dave Richardson, for one, liked to label "the handcuffed volunteers," the millions drafted out of civilian life to serve.

Part Three—Mirror of War

Twenty-One 18 Magazines

Japanese left-behinds in February 1945 were still occasionally killing GI souvenir hunters on the island of Saipan in the Marianas chain of the Western Pacific when Corporal Tom O'Brien put to bed the first issue of *YANK, Published on Saipan.*

For the second time, *YANK* was appearing with a variation on the title, now by its own choice. With *YANK Down Under* circulating among soldiers and Marines farther to the west, *YANK, Published on Saipan* was deep inside Admiral Nimitz's share of the ocean. The discreet suppression of the word "Army" was intended to help sales with sailors. To help further, the edition even added a few Navy stringers as contributors.

The Marianas edition was the 14th in an overseas series that had begun in Britain in November 1942. In 1943, eight more had followed: Puerto Rico in January; Trinidad, Egypt, and Hawaii in June; India in July; Iran and Australia in August; and Panama in September. Three opened up on the Continent in 1944: Naples, Italy, in March, Paris in September, and Strasbourg at the approaches to Germany in November. Three came after Saipan in the final year of the war: Manila in July, Okinawa in August, and Tokyo in September.

In addition to the master edition sold to stateside troops, New York ran off several others, including one for Alaska and others for areas abroad not reached by the out-of-country printing operations.

With at least a few changes in each issue to meet local needs, *YANK*, in effect, was publishing 21 different magazines, and all of it with a staff no larger than a few platoons. To put out the Saipan edition, Corporal O'Brien had the help of three others of the *YANK* staff: photographer Sergeant Dil Ferris, of the Air Force; artist Sergeant Jack Ruge; and another writer, Pfc. Justine Gray, of the Rangers.

The overseas printing operations had been born out of necessity. With shipping scarce and paper heavy, it had been decided early on that the only way to deliver news and features quickly to the distant garrisons was to make up each issue in New York, while airmailing mats and reproduction films to points across the globe where *YANK* crews could do the final editing and then print and distribute. An immense amount of paper was used, with more than 2,000,000 copies a week published at its peak. *YANK Down Under*, for one example, used 770 tons in its first eight months.

The strain on personnel was more dramatic. *YANK* needed EM writers, editors, cameramen, artists, cartoonists, printers, circulation men, promotion specialists, and accountants to make up the paper in New York, to man the overseas offices, and to get the product to every possible camp and foxhole. To do that, Forsberg lamented to the War Department on November 17, 1943, all he had in his *YANK* command were 105 privates, corporals, and sergeants, and four of them no help editorially or in production. Their job was to remind *YANK*'s crew, through occasional close order drills, that they were in the Army.

Forsberg begged a reluctant Washington that what he needed were at least half as many more EM. The colonel had ample horror tales to tell the War Department. Two EM were the total staff of the two Caribbean editions, handling all the editing, printing, and circulation chores. Another two were doing all the news coverage, promotion work, and distribution inside the vast Alaskan theater, a task "impossible, regardless of zealousness." *YDU*, with only two writers and one photographer, had been required to borrow a few helpers from an MP battalion. Panama had no photographer or artist—just two writers who were kept busy with production chores with the result that correspondence was out for them and the Panama Command was shortchanged in the world book. In New Delhi, one EM, with no background in publishing, had to work alone for a single six-week period getting out the India edition. In Cairo, the situation was even more drastic. When General Eisenhower requested an edition for Naples, the small Cairo crew was shipped there, leaving EM of *Stars and Stripes* to do what they could for a *YANK* edition serving all of Africa. The result? "An edition in Cairo inferior editorially and in terms of production," said Forsberg.

Sometimes bad went to worse. *YANK* never missed an issue, but on one occasion, the Indian edition came close when mats failed to arrive. Nonetheless, Sergeant Hargrove managed to get the edition out single-handedly. As was done in other editions, he had made it a practice to save mats that were not used when local material was inserted. At a typical edition, that backup file of mats ran to about 25 pages, just over the size of an issue. Hargrove had received a copy of the current London edition in the mail. Managing somehow to make engravings from pictures clipped from the UK copy, tossing in various of the saved-up mats, and writing new local material, the sergeant got the issue out on time.

In the home office, shorthandedness was just as bad. Although 60 members of the staff—half the total—were posted in New York on a rotating basis, only one did the colonel consider as a competent writer. Despite *YANK*'s lavish use of art, there had been one especially dismal month when every sketch had to be done by the same man.

How the various editions handled circulation was illustrated in Naples when reassigned staffers from Cairo arrived there. Trucks carried copies to the Fifth Army headquarters, where they were broken down into bundles for divisions, then regiments, then battalions, and, finally, shipped to the front lines, along with rations and ammunition. As in the case of all combat areas, the usual five-cent charge was waived and troops got the Sad Sack, the pinups, "Mail Call," and the war reports free.

All editions borrowed help from other military units, an undesirable situation because the extra hands could not be shifted from one area to another, and because such troops were subject to recall at any time. In addition, *YANK* could not reward the temporary staffers with promotions, no matter how great their contributions. Local civilians also were taken on, all of them financed by the nickel charge.

O'Brien's Saipan edition was typical of the overseas operations, except that it was the first time *YANK* was self-sufficient using its own press. Forsberg had arranged for a portable press built to his specifications in Westchester County, New York. It could be moved from one island to another as troops advanced in the Pacific. Elsewhere, local printing plants were used, such as in Paris, where the Wehrmacht publication hastily evacuated the printing plant of the *New York* and *Paris Herald Tribunes,* and *YANK* moved in behind them. Leftover copies of the German publication were put to use as wrapping paper. Forsberg planned three of the mobile presses, but only one other got into use. That was in bomb-shattered Manila.

Ingenuity was needed to make the Saipan operation work. The offset press needed lots of clean water, but Saipan pumps were running dry. By noon, the last drops would dribble from the island's faucets, with no more flow until evening or, sometimes, not even until midnight. Offshore were rusting float tanks, which had been used as piers during the landings. One was dragged up the beach and put to use storing up 1,000 gallons overnight. It was the envy of neighboring troops, until the third day when the rust-weakened sides collapsed under the weight.

Even when water was plentiful, problems ensued. The plates for offset printing needed an albumin coating. Constantly, it wore off, making it necessary to change plates. Corporal Emmet Coggin, an expert on that type of printing, managed at last to come up with an answer. Saipan's volcanic coral earth was tainting the water, and tropical heat was compounding the difficulty. Deep etch plates were needed. *YANK* in New York shipped them and, soon, 70,000 copies a week of *YANK, Published on Saipan* were being shipped to Guam, Tinian, and other West Pacific islands. With EM manning three shifts, the press ran 24 hours, six days a week.

All sorts of extra help were picked up. Checking through what was available at the replacement depot, the "repl depot," a collection point for unattached EM, *YANK*'s officer in charge, Major Jack Craemer, now of San Raphael, Marin County, California, made a happy find. Pfc. Robert Glenn had been a job printer in Florida. He was hired on for more or less permanent service. Less fortuitous was the choice of a husky fellow who seemed ready-made for heavy lifting, such as loading stacks of the magazine for shipment. Craemer noticed that the repl depot seemed eager to turn him over, but only later found out that he was not only unreliable as a worker, but also that he was on parole from a federal penitentiary after having served time as an armed robber. A graves registration team on Iwo Jima needed assistance, and the relieved Craemer shipped the felon there.

With their homes and businesses shattered, natives of the island were herded into a compound at night. Those who wanted to farm or to take jobs were allowed to leave the enclosure in the morning, but the Navy, as island commander and fearing inflation, set 35 cents an hour as the maximum wage. *YANK* used nine young women to assemble and stitch copies and, perhaps on the theory that the Army was not governed by Navy law, paid the allowable third of a dollar, plus an equal amount, under the table. As a result, working for *YANK* became popular.

Along with the women came a youth who seemed not more than nine years old, but must have been a few years beyond that. His language was the native Chamorro but, to go with it, "Jerry" had a little Spanish and some Japanese from previous occupiers of the island. To those, he now added a bit of GI English. He became the unit mascot, turned out to be "a surprisingly effective pressman's helper," and posed in big gloves for a mock boxing match with the tall Craemer. At age 83, the officer still kept that snapshot. Thinking back across the decades to Jerry, Craemer muses that if YANK's old printer's devil were still alive, he must be now a Saipan elder states-man. "I have always regarded him as a genuine fellow alum of *YANK*."

One more source of help was the P.O.W. camp. The Japanese PWs were picked up early each day and locked up again at night. They worked as laborers, although one day the *YANK* team found that at least one of them could be drafted for more specialized assis-tance. He was helping move newsprint rolls from the dock when, in a few words of English, he mentioned that the task was not his first experience with an aspect of the printing business. He had run an offset printing operation aboard a Japanese ship, which harbored occasionally at San Francisco. Pulled out of the labor gang, he was soon helping run the press.

Sometimes there were difficulties with the PWs. "They were muscle-bound, face-conscious little guys," the former major remem-bers. It came to a head one day when they refused to straighten up a couple of 55-gallon drums of printers' ink. Too heavy, they sulked. The episode is clear in Craemer's memory, "I was standing by, watching. Neither the Marine guards nor the *YANK* crew were having any luck with reasonable persuasion."

Craemer had a bad back from a high school football injury, something he had concealed when he applied for a commission, but "I saw what needed to be done to face down those guys. I just walked over and manhandled one of the drums to an upright position," he recalled. "In my back I felt the muscles, tendons, or whatever torn apart. We didn't have any further problems with Jap PWs refusing to work but I sure as hell had back pain for years to come. The first months after the war, while on terminal leave, I was an outpatient at the Letterman General Hospital at the Presidio in San Francisco getting therapy and a brace. The pain continued into my sixties."

There were a few unexpected benefits for a GI assignment to the Saipan operation. On the next knoll was a papaya grove with sweet fruit, a welcome supplement to soldier fare. Offset printing called for an air-conditioned darkroom and a refrigerator for storing chemicals. It was a relief from the oppressive, humid tropical heat to cool off in the darkroom, and there was room in the cold box for beer and sodas. Also, for those who fancied them, there was an abundance of giant frogs and snails. The PWs feasted on them as a delicacy, but *YANK*'s GIs passed them by.

For Craemer, as the officer in charge, only one part of the operation was "none of my business" and the sole reason for all the work, the editorial content. Corporal O'Brien handled the editing and "took his marching orders from McCarthy."

"I was out of the loop," Craemer recalls in a revealing insight into how the officer–EM relationship worked inside an outfit where the EM uniquely had the upper hand in all that counted:

As an enthusiastic reader of *YANK*, I probably made comments favorable and possibly unfavorable on occasion, although I can recall none of either. My take on Tom O'Brien was that he didn't have much use for officers, including me. So we were never close. I regarded that as his problem, not mine. I couldn't see that it made any difference in *YANK* operations, except for a bit of discomfit on my part.

YANK

Liberating 19 Switzerland

YANK's staffers had a problem. They had to be enlisted men to understand the plights of those for whom they spoke, but they also needed freedom to cover the world the way civilian war correspondents did.

Ever supportive and sympathetic, General Marshall listened to pleas from Forsberg and, on May 2, 1944, issued Adjutant General's Order 451.02, spelling out *YANK* rights:

> Editorial personnel of *YANK* will be given the same assistance and general professional consideration as are given war correspondents of civilian publications, consistent with military policy and procedures.

The six last words seemed to leave ample room for local brass to take away much of what the first part of the sentence granted, but the hundred or so *YANK* staffers dispersed around the world already had devised means, sometimes just short of felony, to bridge the gap between their lowly ranks and their unprecedented assignment. Some pretended they were not in the Army at all. Others doffed their American uniforms to put on those of other nations. Photographer Bill Young, for example, dressed as an Australian to cover the Aussie invasion of Borneo. One staffer even "liberated" neutral Switzerland.

Switzerland's "rescuer" was five-stripe Sergeant George "Slim"

Aarons, an eminently self-confident string bean, six feet, four inches tall, who had enlisted in the Regular Army at 18 and had earned the stripes of a technical sergeant as the official photographer of the military academy at West Point. At the Point, Slim got to know, on a first-name basis, the colonels who became the multi-starred generals of World War II. Never, after that, did they hold for him the awe other EM felt. Also at the academy, Slim took official photos of the cadets, including the last image of an occasional suicide. For him, the military mystique was so much "bull," as he often remarked.

At the war's start, the rule at West Point was that no one could transfer from the institution unless he was going immediately abroad. Tapped for *YANK*, he was shipped out at once, becoming one of the first members of the magazine staff to enter a theater of war. Slim headed for Cairo to cover Montgomery's Eighth Army in its counterattack against Field Marshal Erwin Rommel's Afrika Corps. In Egypt, Aarons wangled a jeep, which helped him to become one of the most footloose EMs in North Africa, the Middle East, and the European continent.

Slim was commended regularly in the round-robins, especially on October 13, 1943, when McCarthy recounted the photographer's latest "remarkable stunt." Slim and his accompanying writer, Sergeant Burgess Scott, heard in Cairo that the Free French had captured Nazi spies in Syria and were arranging for a firing squad execution. Making use of their invaluable wheels, the two raced 1,500 miles across mountains, a desert, and three national frontiers, arriving at the execution site in time to give the soldier magazine its first step-by-step report on one of the grimmest aspects of war. Even the coffins were photographed. In pencil on the envelope enclosing the film was a note from Aarons to McCarthy and to Sergeant Leo Hofeller, the makeup editor. "Hiya, Joe and Leo," it read. "How do you like this for a story?"

McCarthy decided to publish all of the grim process, but one element had to be eliminated. With the pictures shown beforehand to the Free French Mission in Washington, *YANK* had to give in to their insistence that the shot of the coffins not be published.

What he covered, Aarons reminisced at age 80, was "the real war, not the movie version." For him, the conflict was a mass of confusion inside which he saw no reason to "play by the rules." "Everything," he said, "was loose. I went where I wanted. I did my job."

While other GIs wore the uniform they were told to put on, Aarons dressed as he pleased. Donning civvies cost some German spies their lives, but Slim tried it on occasion as a way to get out from under vexatious officer control. Changed dress, he found, even opened the way into "officer country," where liquor flowed and life was easier. The sergeant's favorite costume for everyday purposes was a Special Services uniform. "It had lots of pockets to carry things," he recalls.

With the capture of Paris, broad new areas for shenanigans opened. Answering to his own orders, Slim drove into the "city of light," heading for the Rue de Berri and the peacetime office of the *Herald Tribune,* an immediate war correspondent hangout. *Life* magazine photographer Bob Capa was there. "Capa knew all the girls," Slim remembers, so the two teamed up at once to see the sites. Capa steered Aarons to good lodging, the Hotel Scribe, where the charge was $2 a night. Among the women Aarons promptly met was a French writer whose name slips his mind but whose problem stays fresh in his thoughts. She was the catalyst for Slim's and *YANK*'s Swiss adventure.

The young lady's boyfriend was giving her concern. He was Picasso's agent in Switzerland. She and he mourned their separation but war frontiers were keeping them apart. Ever gallant where the weaker sex is concerned, Aarons recalls in his ninth decade that after the war he even managed to "take a girl away from Jack Kennedy," a feat so remarkable that father Joe asked to meet the man who did that. The French lady's plight became Aarons's prompt concern. Contact was made with the lover pining away in Switzerland. The latter came up with good ideas about the best ways to slip through the frontier, and even arranged a cache where the soldier photographer could doff the Special Service uniform, replacing it with inconspicuous civvie shirts and pants. As a tip of a hat to the rules of war, Slim pinned his old West Point insignia to the blouse.

What soldierly logic, beyond care for the frustrated lovers, could Slim cite if his disguise was penetrated? Slim thought of one. Patton was in full flight toward Germany. What was so wrong about him crossing neutral Switzerland to get to Berlin that way?

All was fine, Romeo and Juliet back together, but Aarons ran out of money. In Bern, he asked the American embassy to send a message to Forsberg, requesting $250 for him. Allen Dulles was in the house arranging the Italian capitulation so the embassy communications room was swamped with traffic, but the EM's money appeal was squeezed in. A good news–bad news answer quickly came back. In essence, Forsberg told his staffer, "Here's your two fifty, but if you are in the process of deserting, prepare yourself for 20 years in Leavenworth when this war is over."

Aarons was stunned. It was the least affectionate message he had ever received from Fort Bartholomew. Back in one of his uniforms at *YANK*'s office in Rome, he was shocked anew by a further dispatch from New York. All five stripes were gone. He was back to buck private. He protested, but Major Weeks turned a deaf ear to him. "What did you expect?" Weeks cabled back to him. "A medal?"

Aarons asked for a transfer to the even more free-spirited Air Force, but Weeks refused it. The magazine still had ample use for his talents. Reverting at discharge to his old Military Academy assignment, Aarons was able to leave service once again as a sergeant. As a civilian, he became an eminent fashion photographer.

So far as straying out of proper uniform was concerned, few on

YANK outdid for eclecticism Sergeant John P. "Jay" Barnes, on duty in Chungking, China. Amid uniformly clean-shaven fellow soldiers, he faced the world from behind a huge moustache. McCarthy, in his June 9, 1943, round-robin, said that Barnes's weird mix of uniforms and other things "was an MPs nightmare." Back on a brief home leave, Barnes showed up in the office "wearing one of those tunics that the British issue in India."

Joe described it as

open at the throat and belted with loose skirts that hang down over the waist of his trousers. He has (China–Burma–India) insignia on his left shoulder, an Air Force insignia on his right shoulder, cloth Air Force insignias sewed on his collar and silver gunner's wings on his chest. He has a pair of GI suntan slacks with picturesque grease marks and a pair of high brown suede shoes with sponge rubber soles purchased in India to ease his foot which was folded up like an accordion when his plane crashed.

For a domestic note, McCarthy passed along what Barnes had written to his wife to prepare her for his arrival. He advised her to get a good look at the floor and walls "because (you) sure as hell won't see anything but the ceiling when I'm there."

Aarons's freewheeling in Europe was matched on the other side of the planet by another of *YANK*'s photographers, Sergeant Bill Young, a former cameraman for the *San Francisco Chronicle*. Called early to service as a National Guardsman, he was given the job of personal cameraman for the general commanding the Seventh Division. Very much a rough diamond and an unpretentious man of the people, Young was uncomfortable with the elegants in the general's immediate circle. He was sure, too, that "they didn't like me a bit."

In his heart, Young says, "I was more or less a civilian in an Army so-called uniform." When Larry McManus, a *Chronicle* reporter and a newly inducted writer on *YANK*, suggested that Bill transfer to

YANK, he grabbed the chance. In the invasion of Saipan, Young ran across some of the old Seventh Division officers who were there as observers. It was no happy reunion. In response to the Marshall directive, *YANK* staffers, on that occasion, were being treated as war correspondents, meaning that they, along with the civilian reporters, ate in the field officers' mess.

"I got some looks from (the Seventh Division) officers," Young still remembers with some bitterness. "But they finally gave up when a couple of civilian guys explained the facts of life to them."

Young wandered at will across thousands of watery Pacific miles, sending in photos to keep his *YANK* franchise. He carried no arms. At the invasion of Kwajalein, when every soldier was told to carry a weapon, he refused a submachine gun as too heavy along with his 40 pounds of camera gear. Next, he rejected a lighter carbine rifle. Finally, to achieve peace with a badgering sergeant, he accepted a .45 caliber pistol. Fiddling with it, he discharged it, sending a bullet into the ground between another non-com's legs.

"I gently took the gun and threw it as far as I could out into the waters of the lagoon," he recalls. "Lost in action, it was."

It was a case of adjusting to what fellow photographer Aarons called the often absurd and chaotic real war, not the more purposeful conflict of Hollywood's customary imagining.

"I hardly ever wore a conventional outfit," Young recalled.

On his belt, he would carry two or three canteens of water, except on one occasion on Saipan when the contents were Santori Scotch. Young had found a quart in the remains of the residence of a Japanese admiral.

Along with two camera cases, Young lugged a parachute bag containing sox, underwear, a pillow, silk parachute sheets, and an air mattress. There were also as many cartons of cigarettes as could be squeezed in. No smoker himself, Young found cigarettes to be door openers when favors were needed. After the fall of Japan and the

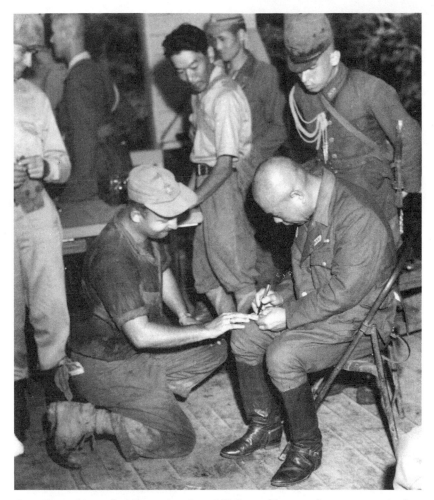

Gen. Yamashita refused to give Sgt. Bill Young his sword as a souvenir, but did sign his "short snorter," a dollar bill with autographs. *(Courtesy of YANK/Bill Young)*

occupation of southern Korea, a roll of toilet paper was added. Dysentery, by then, was widespread. "I would get out of a jeep or a taxi and let fly," the wandering cameraman remembers.

Cigarettes as a form of currency were supplemented with real money when Young and McManus stumbled on not only the admiral's Scotch, but also a straw suitcase bulging with yen on Saipan.

175

Satisfied that it was legitimate war booty, but keeping the yen a secret nonetheless, Young was not sure it was real money until *YANK* started its Tokyo edition and the photographer saw identical 100 and 1,000 notes being run off on an adjacent press. All they needed to become negotiable was an authenticating red chop. All of Bill's notes had the chop. At airfields, many of the bills got him jumped in priority from grade four to two, and to instant takeoff as he flew from island to island.

Rattling loose, the audacity of the *YANK* photographer had few limits. When the "tiger of Malaysia," General Yamashita, was captured on Luzon, Bill reasoned that "he would have no further use for his sword," so, brazenly, he asked for it. "I have to give it to MacArthur," the vanquished general "sort of smiled."

For Young, that was no stopper. "I saw it first," he persisted. "Give MacArthur another." No luck. "He laughed."

Young's scams were not his alone in his area of the war. When he teamed up with other *YANK* staffers in bombed-out Manila, he and the others had food to spare by requisitioning it in the name of such fictitious units such as "the 1687th Prophylactic Medical Unit." There was even enough food left with which to entertain Red Cross nurses.

YANK had officers at various locations to maintain some military discipline among *YANK*'s staffers, but many slipped out from under them while still doing their magazine job. Major Craemer remembers one of the slippers on Saipan. "He was a circulation guy," he said,

You might call him a road salesman. He went by his last name only, no rank. His non-uniform consisted of khaki pants and shirts, which American enlisted men did not wear on Saipan; they wore fatigues. He had no insignia at all. I am sure that most people assumed he was a Navy chief. Actually he was a plain Army private and I thought he aspired to nothing else. He traveled all over the Pacific. I presumed he had some kind of orders but I practiced 'don't ask, don't tell.' Then, when someone did ask how he got to such-and-such distant place from which he had just returned with a big order for *YANK*, his reply usually was that he had run into his old friend, General So and So, who happened to be heading to the same destination and invited him to ride along in the general's plane. We never knew where he was, where he was going, or when he would return, but he always showed up with genuine new orders for *YANK*.

Walter J. Maxwell, who finally made sergeant, no doubt was the major's mystery man.

It was, to be sure, a strange way to run an army, with EM in the field bypassing officers and doing their dealings just with other EM back at Fort Bartholomew, but it did get out the magazine. Ozzie St. George, of the *YANK Down Under* staff, reminisced about the deceased Maxwell in 2000:

He was one of our best circulation men. He came to *YANK*, I believe, from a medical outfit that wished to get rid of him. Before the war he'd sold vacuum cleaners. Many higher ranks on meeting Walter assumed he was a civilian since he certainly didn't look like a soldier. But he was a magician when it came to moving the magazines. By July '45, the *YANK Down Under* office was still in Sydney but the front was in the Philippines.

It was like publishing in Miami for delivery in Alaska, a daunting circulation challenge with which only a Maxwell could cope. Possibly, the enigmatic soldier was reporting as one EM to another in Sydney or New York, leaving officers like Craemer out of the picture. It was a strange way to run an army but, as *YANK*'s circulation soared into the millions, it was evident that it was a way of doing things that worked.

Drawing 20 a War

YANK's mature contribution to the war effort began when it sent its artists, photographers, and writers into combat. That was when the GIs on 42nd Street truly realized they were in the Army, giving up a basically civilian life in which they had enjoyed an adequate living allowance, had slept in hotels rather than barracks, had no use for mess kits as they dined in restaurants or ate 25-cent meals in the nearby Automat, and had shuffled across the sawdust-covered floors of Tim Costello's bar for an evening beer.

First came deployment around the planet, which was not each GI's choice. In that, there was officer control.

"I don't recall ever being consulted on the matter of assignments," artist Greenhalgh thought back at age 85. "Whether I wanted to go out in the Atlantic on a destroyer or to the South Pacific. We didn't volunteer. They jolly well just sent us."

Forsberg, at 95, mentioned who "they" were. "Jack Weeks (a major, the executive officer) and I would talk it over with McCarthy and we would let Joe make the announcement."

To cover global combat, the magazine had hardly a dozen artists. They had been picked up haphazardly, some already embarked on careers as commercial illustrators when they were drafted, others just out of art school. Scooped into the Army, most of them were told to put away the effete brushes and to shoulder carbine rifles.

Greenhalgh had just begun a freelance career, doing illustrations for the *Coronet* and *Esquire* magazines and for ads for Chicago's Marshall Fields department store. Sent to Camp Wolters in Texas, he became a drill sergeant. Private David Shaw of Boonville, Indiana, drafted from an art studio in Chicago, was assigned to paint labels on crates in Naples. Howard Brodie, an exception, was invited to join *YANK* by enlisting voluntarily, quitting his job as a sports artist for the *San Francisco Chronicle*, and skipping basic training. Even so, his ride east on a troop train gave him some of the flavor of military life. He recalled that he learned from soldierly exchanges that there was a multipurpose, four-letter word suitable for service as a verb, "—— you," as an adjective, "this ——ing war," and, as a noun, "what the ——."

The road to *YANK* was different for each of the artists.

Greenhalgh, busy most of the time shouting "left face" and "right face" as a drillmaster, amused himself in spare moments sketching camp life. As an afterthought, he mailed the pictures to *Coronet*, which published them in a full-color foldout. His camp colleagues, he remembered at 85, stared at him with new respect when, on maneuvers on a Texas range, he received an order from Secretary Stimson shipping him to *YANK*. EM and officers alike crowded around to stare at the order from Washington.

Shaw, in Italy, came across an art school contemporary who was working as an illustrator for the Neapolitan edition of the *Stars and Stripes*, and he introduced him to Sergeant Harry A. Sions, editor of *YANK*'s Naples edition. Sions arranged for a transfer.

Joe Stefanelli, just out of art school, was employed in what he described as "a boat and shore outfit" when Sergeant Doug Borgstedt heard of him and got him moved to *YANK*.

The artists had one advantage over the photographers as *YANK*'s illustrators. Something might happen so fast a cameraman could not focus in time. Photographers also needed lighting and, under pressure, had problems with composition. Neither difficulty affected the artists. They could take mental notes, creating their pictures later, placing emphasis where they chose.

One thing photographers and conscientious artists did share was the need to be exposed to front line fire. Timid writers sometimes had an alternative. They could pick up copy secondhand in the rear.

Reminiscing later, several *YANK* artists admitted how terrified they often felt. Brodie said that on Guadalcanal, for *YANK*'s first staff-illustrated combat report, "I searched for bodies to dull my dread of the dead." Ill with malaria and hepatitis, he was invalided home, temporarily, to an Army hospital on New York's Staten Island, his nerves so jagged from the memory of air raids that every time he heard a siren, the Guadalcanal warning system, "I felt like dropping to the floor, still fearing Washing Machine Charlie."

Shaw said he had been frightened even by the mock explosions in basic training exercises. On the static 1944 Apennine front, between Florence and Bologna, his trembling under incoming and outgoing artillery fire made his mattressless bedsprings squeak. The mattresses had been removed to a bomb shelter for use down there. He was with the headquarters company of the 91st Division, set up inside an astronomical observatory. Shaw noticed that others, already adjusted to the sound of cannonading, were sleeping peacefully. The shelling was just the routine late-evening exchanges between the American and German artilleries. Out of that trip to the front, Shaw got two spreads for the world book and a cover for the Naples issue.

Greenhalgh too had enduring memories of being "scared to death."

There was a debate among the *YANK* artists about the ethics of combat sketching, and whether there was any such thing as a battle painting "done under fire." On that, Greenhalgh had strong feelings.

"Any artist," he said, "can sit back at his drawing board somewhere and draw any darn thing. He can draw 'a war.' But the way I feel, the artist has a duty to draw what he actually sees. If he doesn't see it, he is not entitled to draw it."

With that as his conviction, Greenhalgh's paintings of a carrier raid on Wake Island and combat in the Solomon Islands and on Guam had such a ring of truth that he earned a coveted award of merit at the 1944 23rd annual Art Directors' exhibition. The award was for a full-page study in *YANK* of primitive GI housing in the Solomons.

As for whether a sketch of close combat ever, in honesty, can be captioned "drawn at the scene," Greenhalgh thought not. In one gunfight on Bougainville, he said, "I tried to make a drawing; I was worried and jumpy and nervous and kind of hunkering down to get away from whatever it was, a tank going this way and that." He still had the pad on which he had tried to record the scene. There was just a scribble. Even he no longer could figure out what it was that he had tried to record. Working afterward from memory, he got a published picture out of it, but it was not labeled "done at the scene."

Perhaps there are exceptions to all rules. Greenhalgh recalled his astonishment in one pitched battle on Guam. Atop a telephone pole, a Marine painter was recording the fight.

Brodie, too, as a commercial combat artist after World War II covering French Indochina, Vietnam, and Korea, found that the on-the-scene feat could be accomplished on occasion. Under fire in Korea, he said, he painted one of his lifetime's finest studies of the taut face of a combatant. Midway in doing the sketch, he had a second thought: "Geez, I better be careful or I'm going to get shot."

There were times when *YANK* artists were in danger from fellow Americans, other occasions when their work could not be published, and still others when higher reasons caused them to put down their brushes.

In the chaos of the reconquest of Manila, an American soldier aimed his rifle at artist Stefanelli, mistaking him for the enemy and shouting to him to raise his hands in surrender. "Fortunately," Joe said of his challenger, "he was not trigger happy, and I yelled frantically 'I'm a *YANK* correspondent.'"

Joe survived to become a member of the New York school of abstract expressionists, with 40 one-man shows in New York, Philadelphia, Princeton University, Rome, and Hamburg, one of them as late as 1997. At the Cedar Bar in Greenwich Village in the 1950s, the yet unknown, and often threadbare, painters of the "New York school" were his buddies—DeKooning, Pollack, Kline, and Guston. Stefanelli's regret was that his paintings for *YANK* on Mindanao and Luzon in the Philippines were done at 23, when his style was still unformed.

At the Battle of the Bulge, in the death throes of the Wehrmacht, Brodie had an experience at the hands of fellow Americans similar to that of Stefanelli. Young Germans in American uniforms had infiltrated as spies. Brodie became suspect as one of them.

"Who pitched for the Yankees?" a ragged-nerved GI tested his knowledge of things American.

A San Franciscan, Brodie wracked his brains for New York baseball lore. "Answer for Christ's sake," the rifleman demanded. "I'm a *YANK* artist," Brodie pleaded just as Stefanelli had done. "Then, who's the best artist on *YANK*?" Humbly omitting himself, Brodie tried "Sad Sack." That did it and the artist was able to breathe easier.

It was the very infiltrators who gave Brodie one of his most painful sketches of the war, and one he was forbidden to publish until the conflict was over. Three trembling youths, shivering in American warm-weather cotton suntans, were tied to stakes. White discs were pinned above their hearts. MPs in heavy winter dress took aim.

"Sadness surged within me," Brodie remembered at 88. "I stifled a feeling to cry out. I swore in the immanence of those deaths to fight for future lives."

Sgt. Howard Brodie's sketch of a spy's execution, still marked with the census stamp forbidding wartime publication. *(Courtesy of* YANK/*Howard Brodie)*

Brodie's fearsome painting showed a dead boy slumping forward, blood gushing from his mouth. The censor, fearing reprisals, banned publication until after the war. The executions took place inside the 82nd Airborne Division section of the front. Soon afterward, back in the Ninth Army press camp, German artillery struck the building to which Brodie had retreated. The ceiling collapsed. A soldier's leg was torn off. A correspondent was killed. Even so, said Brodie, "it didn't bother me as much as the execution, three young men lashed to stakes and, calculatedly, shot to death," somehow so much worse than "the random chance of death in battle."

As the war dropped in 1944, Brodie's paintings reflected increasing compassion for the sufferers he sketched. One of Brodie's paintings for *YANK* showed a soldier with arms around a comrade whose terror had overcome him. It was in a farmhouse, with the shells of German 88s crashing in from one side and the 75s of American tanks battering from the other.

"I remember how he screamed," Brodie said of the soldier. "He was just out of control and he screamed. The other soldier embraced him, consoling him. That was a moving moment for me to see such compassion in combat."

Soldier artists had two ways to judge one another: the professional quality and impact of their work, and the courage to go into combat action.

The hands-on winner in the first category in World War II was the Pulitzer Prize–winning cartoonist Sergeant Bill Mauldin, whose long-suffering Willy and Joe epitomized the GI plight. Where it came to courage, however, Mauldin tipped his hat to Brodie.

Former *YANK* Corporal Tom Shehan remembers the relationship between the two artists. Mauldin had been a *YANK* cartoonist for its first few issues, but he concentrated later on *Stars and Stripes*, where his fame was made. He remained close to *YANK* staffers, however, his Naples studio one of their hangouts. Shehan recalls a comment Mauldin made about meeting Brodie at a time when an American unit was attempting to cross a river under German fire.

"It was too hot up there for me, and I was on my way back to wait until they cooled Jerry off," Shehan quoted Mauldin. Strangers until then, Brodie was effusive in congratulating Mauldin on his fictional characters. Later, the river crossed, and the two of them on the other side, Mauldin asked when Brodie had made it over. With the first wave of assault boats, he was told. The cartoonist confessed his awe.

Though Mauldin was generous in his praise of Brodie, he had no reason for shame. He was awarded two Purple Hearts for wounds.

At one point in the Battle of the Bulge, Brodie put aside his pencils to help the medics. For his efforts that day, he was awarded the Bronze Star for heroism.

Sergeant Merle Miller, editor of the Paris edition, taunted Brodie as "a chaplain," a bleeding heart, but when Brodie objected to an editorial Miller had drafted, the editor backed down. Miller, making full use of the freedom he had been given as an EM, had planned to publish, in what was an official War Department publication, "If you meet a pregnant Nazi woman, kick her in the belly."

"*YANK* editorials reflect the views of all of us on the staff," Brodie protested. "And those views are not mine."

No such vile editorial appeared.

Were there some scenes best not sketched? Greenhalgh thought so and cited one. It was on the carrier raid on Wake Island. One returning plane crashed into others on the flight deck.

"It threw gasoline all over the place," the *YANK* artist remembered. "Eight or nine men were on fire. They were torches, running all over the place."

One sailor was sprawled aflame on the deck, "his face as white as if it had been burned up."

A photographer moved in to take his picture, and the dying man waved him away saying "don't."

It was the grim reality of war, which *YANK* normally shared unsparingly with its serviceman audience, but Greenhalgh never drew it.

"I don't know why," the octogenarian revisited it in memory.

"I guess it's because … I don't think that kind of intimate gore is too good."

It was the sergeant's first encounter with the rank-rigorous Navy. Taught how to behave as he climbed the gangplank, he waited until the officer of the deck said, "Welcome aboard," and then responded crisply, "Glad to be aboard, sir." Escorted to "an endless series of rooms" in officer country, he was taken out again five minutes later when his papers were examined. Shifted to a berth among the chief petty officers, he found compensations. "It was right beautiful for an enlisted man, great breakfasts of steak, eggs, toast, hash browns, grits, orange juice, fresh grapefruit, apple pie with ice cream, milk, and coffee, most of it all at once."

Each artist had his own style—Greenhalgh's sketchy and loose, Brodie's completely fleshed out, Stefanelli's photographic with no hint yet of his future abstractions, and Shaw's quick, light, informal, and impressionistic. As one of the highest paid editorial and advertising illustrators in the 1950s, Shaw became known as capable of instant production. *American Artist* in 1958 said of him that the first thing he did on getting an assignment was to phone for a messenger. By the time the carrier arrived, the work would be ready.

YANK

Closer Than 21 Artillery Range

Even more than *YANK*'s artists, *YANK*'s corps of 20 photographers were exposed to battlefield dangers. A double goal was to get as close as possible to front line action while living long enough to record it.

Sergeants Dick Hanley and Dave Richardson, in the MacArthur area of the Southwest Pacific, made a joke of it: which of them could get closest to the enemy while surviving to get the picture back to the magazine?

Young Richardson, who had been a copyboy on the *New York Herald Tribune* before getting into the Army, had the better of it in the final days of 1943 when his picture of riflemen in action in the battle of Sanananda was picked up from *YANK* and reprinted in papers in New York and other cities across America. Only 30 yards separated Richardson's soldiers from the Japanese.

Then it was Hanley's turn. With the Marines at Cape Gloucester, he shot pictures of combatants only 18 yards apart.

Hanley wrote later to New York about the "queer" way that coverage had happened. He and sergeant writer Ozzie St. George were doing interviews at a Marine battalion command post when they heard of "an outfit which had refused to be relieved from its front line position out of revenge for a dead mate who had ended up on the wrong side of a sniper's bullet." The command post itself was a lesser story; it was 50 yards from the close quarter combat.

"These Marines don't stand on the formal rules of the book about how far a CP (command post) should be from the front," Hanley noted.

The two from *YANK* decided to go see the Marines who would not pull back. They were told to look for Sergeant Barkauskas, who was in charge. Up a narrow jungle trail the two went, until they met a husky Marine carrying "two huge mess kits full of chow," food for the determined fighters. He was Barkauskas. The three joined forces.

"All the time, the firing of machine guns, tommy guns, rifles, and mortars got more furious and closer," Hanley went on. "Finally the sergeant ducked into a machine gun pit with Ozzie and I right behind him. St. George began to ask questions and I began to shoot pix of a marine in the next foxhole. After a while, we inquired how close we were to the Japs. 'Right out there,' said a machine gunner. 'Right out where?' asked St. George. 'Over there behind those two trees,' said the machine gunner."

They were 54 feet away.

"Oz and I," Hanley wrote, "tried to get our Adam's Apple back into their proper spot."

"Inasmuch as I was up there," Hanley continued, "I decided to try and do the job right. To get the proper angle for a picture of the men with their .30 caliber machine gun, I had to crawl out in front of them with my back to the enemy. The Marines said it was okay for me to be in that exposed position. They said they'd cover me. 'If they get you,' they said, 'we'll get them.'" With a gulp, Hanley took the shot and crawled back.

"These Guadalcanal veterans are so calm under fire that you are not afraid when you are with them," the photographer gave the boys on 42nd Street as his conclusion. Then he modified the thought, "At least, not much."

An irony, and one of the problems frustrating the combat cameramen, was, as McCarthy related in a round-robin, that "the pictures Dick made at that advanced spot look like shots of Louisiana maneuvers." To see them, he wrote, "you would never believe that there was a Jap within 18 miles of the Marines." With troops on both sides hunkering down, it was hard for the camera to record the peril.

For the soldiers and Marines in the Pacific, every landing was an exciting and often fateful adventure. Hanley set what was probably a *YANK* record. He went in on the assault waves in almost every landing in New Guinea and the Philippines, 12 in all. He was one of a half dozen or so staffers to receive the high award of the Legion of Merit.

The moment of fright Hanley and St. George suffered on Cape Gloucester was multiplied many times over for fellow *YANK* photographers both in the Pacific and in Europe. Sergeant Bill Young, in his 80s, remembered the bad moments as well as the good ones, when he and writer Sergeant Larry McManus covered the invasion of Saipan in the Pacific Marianas. The good part was the 14-day ride from Honolulu aboard the troopship *Frederick Funston*. The amiable captain offered to ignore their stripes, putting them in with the junior officers. Knowing, however, how well chief petty officers dine, the two asked instead to bed with them. It was a good decision. For a fortnight it was "the real good life," Young remembers. "Every day, at 3:30 P.M., the bakery would finish the daily bread detail and a petty officer would bring to the chiefs' mess about 20 loaves of very hot bread. A couple of pounds of butter would be laid on the table. The hot bread would be split open, the dough would be scooped out, and then you would start with top or bottom loaded with butter. And there would be a very large glass of milk. That way I gained about 12 to 15 pounds on the trip west."

There was no need to worry about an expanding waistline, however, because "I lost all the poundage on the very first day on the beach."

Wisely led troops moved quickly inland to get away from mortars trained on the water's edge, but the two *YANK* men, answering to no orders but their own, plumped down on the sand as soon as they reached dry ground. There were explosions around them. "We lay about five or six feet from each other," says Young. "There was no cover. I was mostly thinking of the best way to lie there, lie straight out or curl up in the prenatal position. I never came to any conclusion. Each way was as bad as the other."

Later, the two crawled in search of others. "I had nothing I could use for a picture. I needed people." After what seemed like an hour, the two came upon Marines. The leathernecks had horrifying stories. "A Colonel Crow—I think he was a battalion commander—got hit by a grenade out in the water. It blew off his testicles and the rest of him. He continued to stand in the water waist high and, with his swagger stick, waved those poor bastards on, yelling 'you can't live forever' or something like that."

The two from *YANK* bedded down amidst the brain-piercing stench of dead Japanese. "The smell was all around us. You smell it and you don't like it and you can't get away from it either." Next morning, Young learned why. He had passed the night lying on a corpse that had been scantily covered with sand. "McManus got a big kick out of it," Young recalls. "It's funny now but it wasn't funny until later that day."

A rumor added to the tension. The invasion fleet had left, prompting panicky gossip and the false impression that the battle was lost and the troops had been abandoned. "Most of the guys around us were as scared as we were," the terrified *YANK* sergeant recalls.

With one invasion following another, "I really got jumpy and gun-shy," Young relates. "At a movie on Leyte, I heard a sharp crack and, before I knew it, I was under the coconut log I had been sitting on. I needed a vacation bad."

Using his *YANK* freedoms, Young treated himself to a feature assignment, flying with Australians into the high interior of New

Guinea. "We got to the real natives," he says. "Bones through their collective noses. White and black paint all over them. Tree bark covering their penises. Great stuff." There was time even for some conscienceless trading, chewing gum and cigarettes for valuable Stone Age clubs.

Suddenly, the war was over. Sipping pink gin in an Australian officers' mess in New Guinea, Young heard the news.

"I staggered out and threw my glass at a tree," the cameraman says. "I yelled, 'I'll be goddamned, I'm alive at the end of it.' I just couldn't hardly believe it."

Young had had his share of *YANK* escapades, but he had also done his part in fighting the war. He was awarded the Bronze Star for bravery.

The difficulty Hanley had in photographing combat at Cape Gloucester was shared in mid-Pacific by Sergeant Johnnie Bushemi. Being in a battle is not enough when so much is invisible. Bushemi was anguished. After covering the invasion of the Gilbert Islands, he wrote to the home office that "if I don't do better the next time, I am going to quit."

McCarthy took it seriously and, in a January 1944 letter to the field, lamented that Bushemi did not understand how well he was doing:

My gawd, no less an authority than Frank Capra (of Hollywood) thought our John did a hell of a job on the Gilbert invasion. We gave the Motion Picture Branch (of the Army's information division) the colored movies that John took on the transport and at Makin and Tarawa to be included in the *GI Newsreel*, the Army and Navy Screen Magazine. We ... saw them the other day and they are really honeys. John has developed into an expert movie cameraman.

Bushemi, not risking his life for nothing, was urged to keep taking pictures.

While Bushemi had his frustrations covering island warfare, Slim Aarons, in Europe, envied him whatever opportunities he had for close-up picture taking. In his view, Bushemi's fields were greener than his. Although Aarons sent in shocking photos of such battlefields as the all-but-obliterated town of Cassino in Italy, he wrote apologetically to photo editor Hofeller, "I'm sorry I can't give you the going-over-the-top stuff that comes out of the Pacific, but the way they fight over here they never get closer than artillery range to each other except for patrol action at night."

Slim made up for it by imaginative driving in his jeep, not only inside the forbidden boundaries of Switzerland, but all across North Africa, the Middle East, Italy, and France. In addition to covering the rout of the Afrika Corps in North Africa and the Anzio landings, he did features such as one in newly freed Rome where he used 16-year-old English-speaking Jean Govoni as a model.

Decades later, it was still a topic of conversation for Signorina Govoni, now Signora Salvatore, and public relations director for the luxurious Villa d'Este hotel on Lake Como.

"I was so young," she says. "And I had nothing nice to wear. Being a redhead, my mother said I would have to wear green. We had a tablecloth that color so my mother cut it up and made a dress out of it. I had just seen the German soldiers driving away up Via Salaria. They were children. They had not yet begun to shave.

"Mr. Aarons took me to Mussolini's balcony in Piazza Venezia. He had me pose on it tossing biscuits down to the crowds. Mussolini's balcony! And I so young."

Censors, as well as the enemy, caused the *YANK* photographers difficulties. In one painful example, MacArthur's first seaborn invasion, the seizure of Woodlark Island, the censor suppressed 20 of Dave Richardson's 24 photos.

One of the most poignant *YANK* photos of the war narrowly escaped death at the hands of the censors. It showed two exhausted soldiers in a field hospital in Germany. Infantrymen, they had just been recalled to the rear. Pfc. Joseph F. Ieradi hunched forward on a

Pvt. Pat Coffey's portrait of an exhausted soldier. In the upper left-hand corner is the bottom of a sacred painting, indicating that the Army was making use of a church, something the censor wanted to hide. *(Courtesy of* YANK/ *Pat Coffey)*

195

stool, his black hair matted, a two-week growth of stubble on the cheeks, his boots befouled, and his uniform bedraggled. Beside him, lips pursed tight, lay prone another exhausted GI. The shot was one of a series taken by Private Pat Coffey. Forty of his pictures were stricken with the Army censor's red cross, the reason being that the 80th Field Hospital of the Third Army, hard-pressed to find a place in which to care for the wounded, had violated the Articles of War by commandeering a church. Crosses, statues, and religious paintings held prominence in the banned shots.

As editor, Sergeant Ecker had asked Coffey to get a photo feature on a unit withdrawn from the front for "rest and relaxation." The not-yet-disillusioned Coffey had sent back early word to save two or three pages for a "wonderful" layout. Ecker set the space aside but, after all the kills, had only one good shot with which to work. He gave it one full page. Headlined "Portrait of a Tired Soldier," it was still compromising, especially when so enlarged. Just above the pfc's head was the bottom of a painting, someone, barelegged, kneeling before a fully cassocked figure, presumably a depiction of one of the miracles of Christ. It gave away the religious setting. With all his zeal, the censor had missed.

In 2002, Ecker, a New York lawyer, spoke of the effect the picture had. Ieradi, he said, "could have been Everyman—Everyman who has been in the line, that is. My caption began: 'Weariness is etched in every line of his body.' Pat had not bothered to get the name of the second soldier who was slumped on the floor. The photo was picked up by AP and widely reproduced with a teaser caption seeking the identity of the unknown soldier. Five or six mothers in different states claimed him as theirs. Pat gave Elizabeth (Ecker's wife) and me an autographed copy and it hangs on my wall."

YANK's cameramen even had a near miss on the greatest photo to come out of World War II, the flag raising on Mount Suribachi on Iwo Jima. *YANK*'s Pfc. George F. Burns Jr. stood side by side with

Closer Than Artillery Range

AP's Joe Rosenthal as six Marines worked twice to raise a premature victory flag over an island where 25,000 Americans eventually were killed or wounded and a Japanese garrison of 220,000 was all but annihilated. First, a small emblem went up, then a bigger and famous one. The two took shot after shot. Burns's effort was delayed in a Signal Corps laboratory while Rosenthal was notified that AP was delighted with one of the many exposures he had made. He had no idea which it was, but any visitor to the national cemetery in Arlington, Virginia, can see it now duplicated in a 110-foot-high, 100-ton bronze sculpture, six servicemen straining forward, the great flag strung out behind them at the midpoint of the raising. It was the magical moment, a portrayal of victorious struggle. Immortalized in film and bronze, two of the Marines died later on the island, and a third was wounded fatally.

Burns did get an exclusive war shot later, Premier Tojo's attempted suicide. The Japanese government leader shot himself, and Burns helped U.S. Army officers carry him to a bed where, as the photographer noted, the light was better for his purposes. Like Sergeant Young with General Yamashita, Burns had his mind not only on the pictures but also on the premier's ceremonial weapon, a prize souvenir. Unlike Young, he did not ask for it. In the spirit of shenanigans, he slipped it into his trousers. On his way out of the house, the short-statured pfc was so stiff-legged that an MP noticed. He confiscated the trophy, but *YANK*, at least, had a photo spread.

George became a postwar contributor to *Life*, *Look*, and the *Saturday Evening Post*, and founded Burns Photography Inc. in Schenectady, New York. He died of a heart attack at 71, in 1988.

For all of the problems the *YANK* photographers faced—lugging heavy equipment, confronting dangers—there were compensations, and Chief Petty Officer Pawlak found one as the war ended. Thousands of soldiers eligible for immediate discharge were scattered around the world, and there was far too little transportation to get all of them home at once. In Japan, Pawlak gave himself an assignment, a photo essay on how GIs were making their way back. He had no idea whether the layout would ever see print, but that didn't at all bother him.

One way home was in a 30-day crossing aboard a crowded troopship, but Pawlak had his eye on something faster and more comfortable. Yokohama Airport had been shot up badly, but American military planes were using it. Official orders were needed there and Pawlak had none, but he did have something just as good. Rare was the pilot who would pass up the chance of getting his photo in *YANK*.

It was a con, to be sure, but it was also a case of an EM making do with what assets he had. At once, there was a difficulty. Not all the planes were heading east. Pawlak got on just the same, even though some were flying back toward the Asian mainland. That was no matter. There might be a better connection wherever the plane landed. Sometimes, there was not. Pawlak thinks back now to several "small specks" in the ocean where there were no immediate connecting flights. Zigzagging willy-nilly across the Pacific, the cameraman landed on Okinawa, Iwo Jima, the Philippines, the Palaus, Biak, Saipan, Guam, Tinian, Kwajalein, Johnston, Hawaii, the United States West Coast, and, finally, Texas, Kansas, Ohio, New York, and release from service.

None of the pictures were used.

All My 22 Writers

Harold Ross, the editor of *The New Yorker* and a fan of *YANK*, was chided one day for letting a certain article run too long. In one of his round-robins, McCarthy summed up what Ross had to say about that, "Of course it was too long, but I have to fill space. All my writers are on *YANK*."

Not all, to be sure, but there were several and, with them, a mixed bag of others from a dozen publications, some of them writers of experience, others fresh off a newspaper's copyboy bench.

"We must be turning out a pretty good quality of writing," McCarthy added in his April 19, 1943, letter to the staff, for the *Atlantic Monthly* was asking for permission to run reprints. The publication's associate editor Charles Morton had said of the soldier weekly, "There's not a dead paragraph in the book."

A few months later, McCarthy shared further news suggesting that his staffers were finding their way. The *Saturday Review of Literature* cancelled its annual award for distinguished service to American letters, replacing it on a one-time basis with an encomium to *YANK* for "distinguished service to American publishing."

We do this, said the *SRL*, "to call attention to a publishing feat of such dimensions as to be without parallel or precedent, so far as we know."

In an editorial, the *Review* hailed *YANK* as "an intelligent and interesting publication bearing on every page the polished professional touch. … It is readable, likable, and gets across a good picture of the news without forgetting that a big part of its job is to amuse and entertain. If there is such a thing as a morale builder this is it."

"Thus our award to *YANK*," *SRL* said. "In making it, the editors would also call *YANK* to the attention of the Pulitzer Prize Committee as a deserving candidate for one of the journalism awards—possibly the award for public service. Or if no category exists now in which *YANK* might qualify, let the committee invent one."

Tris Coffin, CBS's Washington commentator, wrote much the same to Merle Miller, "*YANK* is damned sound reading, much better than the drivel we get in the commercial magazines. *YANK*, to my mind, is the best magazine of the generation. It is adult, honest, straightforward, and it tells a story."

Coffin, in McCarthy's view, overstated it. "There are quite a few magazines in this generation that have an edge on *YANK*," Joe told fellow staffers, "a big edge in fact," but he was pleased with the kudos. Forsberg, just before his passing in 2002, argued that Coffin was on target. A reason, he said, was that *YANK*'s people, as soldiers, made it a practice to go closer to combat than did many of the civilian reporters.

Looking back across half a century, some of *YANK*'s surviving writers shared memories of their experiences writing articles for the magazine. Few had as many hair-raising adventures as Technical Sergeant Dave Richardson, who is now a resident of Washington, D.C. A lanky, self-described bean pole, weighing 123 pounds, 15 pounds underweight, Dave was adjudged physically unfit for service when a doctor examined him in a New Jersey armory. As a 1940 journalism honors graduate of Indiana University, however, he had the advanced education the Army valued. He was given a choice, excused from service as a 4F or induction as a private. He had just been given a fond farewell from his $15 a week office boy job at the

New York Herald Tribune and "did not relish creeping back to that gang," so he opted for 1A. Soon, inside the Army, he made it known that he wanted to cover combat as a soldier journalist.

YANK correspondents in combat were issued carbine rifles, .45 caliber pistols, and, sometimes, hand grenades, but few made use of them. Dave, on the other hand, soon astonished 42nd Street stay-at-homes by not only lugging weapons, but actually employing them. To get stories in New Guinea, and elsewhere in the Southwest Pacific and in Burma, he faked his way into combat by promising to fire weapons about which he knew nothing. On a B-24 bomber raid against Japanese shipping in the Guadalcanal Slot, he signed on as a machine gunner. Never having touched such a weapon, he learned enough en route to take a shot at an attacking Zero fighter.

Next, on a PT mission in the Bismarck Sea against a flotilla of Japanese supply boats, Dave made his first acquaintance with the BAR, the Browning automatic rifle. It was heavier than he liked, but he managed to fire some rounds.

In a Southeast Asian jungle, Dave was the bow gunner in the first tank in a column. That renewed his acquaintance with a machine gun. He was told to look for orange flashes, the signature of an antitank gun. Firing tracers at one such flare, Dave guided cannons to their objective. It was not easy. "I could not keep my eyes from blinking as hot shell cases spewed forth from my gun. I hollered, swung the gun, and fired a long burst, hoping blindly to be somewhere close. From then on, everything seemed a blur."

It worked. "Nice work, sergeant," the commander, Colonel Rothwell Brown, said the next day. "You helped us wipe out an anti-tank gun."

In another feat, Richardson joined four men of the Office of Strategic Services in parachuting behind Japanese lines in the Burma jungle to teach friendly tribesmen how to use rocket-launching bazookas. This time, it was a double fib. Richardson knew nothing

Courageous Sgt. Dave Richardson. *(Courtesy of Dave Richardson)*

about bazookas and never had jumped from a plane. Bazookas were fired from the shoulder and had impact enough to knock out a tank. They also had a powerful kickback. Fellow GIs showed Dave how to fire one "without blowing off my shoulder."

In flight, the reporter asked instructions on how to jump. "If you see a tree coming," he was advised, "pull the parachute cord to a side. When you land, pull your arms in and bend your knees. Good luck."

On the way down, Dave yanked the cord but landed in a tree just the same.

In his most spectacular feat, Richardson walked more than 500 miles through the tangled Burmese outback with a team of volunteers led by Brigadier General Frank D. Merrill. Sabotaging behind enemy lines was the mission of "Merrill's Marauders." "It was hairy and exhausting," Richardson wrote the office. "Besides the regular 40-pound horseshoe-type pack, which every Marauder carried (containing a blanket, poncho, kukri knife, four to five days' K-rations, entrenching shovel, water wings for swimming rivers, and an extra pair of shoes), I lugged two cameras, film, notebooks, maps, pencils, and my carbine. And I brought along a baby Hermes typewriter."

The one chance to use the writing machine was during a two-day rest following a gunfight. The first day was given to repairs on the typewriter. "Incessant rains" and the jangling on the back of a mule had thrown it out of kilter. Richardson doubled as writer and cameraman, so the second day went to typing photo captions. Except for note taking, the writing of stories was left to later.

YANK broke its rule of one story to an organization, giving Richardson space for a series on the Marauders. For decades after the war, Dave attended the outfit's reunions.

Disease, as well as gunfire, threatened *YANK*'s combat correspondents and Richardson got a dose of both. Already underweight, he lost 30 pounds, slipping down under the 100 mark, only 15 pounds or so for each foot of his height. He caught malaria, had bacillary dysentery, swamp rot, yellow jaundice, and chigger bites, needed intravenous feeding after one spell of jungle coverage, and, for a while, wore a white patch over his right eye after it was nicked by shrapnel from "a big baby that landed only six feet from my foxhole."

Richardson was awarded the high honor of the Legion of Merit for firsthand combat reporting, which made "a substantial contribution to the morale of troops in the Southwest Pacific area." He was given the Bronze Star for combat courage and the Combat Infantry Badge for his walk with the Marauders. He and two other *YANK* staffers, Sergeant Georg Meyers and, posthumously, Johnnie Bushemi, were among the 12 journalists given the 1944 Headliners Award for valor.

Reflecting on how it was to use weapons as well as a typewriter and camera, Richardson saw an advantage. Keeping busy with guns, he said, left less time to be terrified.

Georg Meyers, with whom Richardson shared the Headliners valor award, wrote for *YANK* both in Alaska and in Berlin. Undetected inside a sleeping bag on Attu in the Aleutians, he survived a last-ditch suicidal bayonet banzai charge. Fighting on the island in April and May in 1943 had gone on for 19 days. The Alaskan soldier newspaper, *Kodiak Bear*, praised Meyers for the way he comported himself in a battle that left only 11 prisoner survivors out of a Japanese force of 2,649, while Americans suffered 500 dead and 1,100 wounded.

Meyers, said the *Bear*, "is like his front name, a slightly abbreviated gent, with a mild flare for the spectacular and a liking for thrift in words. He took a chance going to the Attu and Kiska front lines for he is just nicely Jap size and speaks a few words of Japanese. But there the similarity ceases. He is doing a sharp, honest, thoughtful job of reporting."

Georg, too, won the Bronze Star, a recognition of his actions in the Aleutians. But, he says now, the two Attu weeks of "gunfire virtually without interruption" paled in comparison to the close call he had in Berlin 15 months after that. Although Meyers had an arm patch authorizing him to take pictures anywhere inside the fallen capital of Germany, he was accosted at the Alexanderplatz, in the heart of the Soviet sector, by "a wild-eyed Russian soldier." The Red Army EM ran down the steps of "a bomb-pocked stone building," pointing his tommy gun and grabbing for Meyers's camera. Strapped to Meyers's wrist, it did not come loose.

"Interpreting that as a struggle, he jammed the gun into my

midsection. The next sound I expected to hear was 'Taps.'"

The camera finally snapped off and the Soviet smashed it.

"Brandishing the tommy gun," Meyers remembers vividly, "he waved me away. I decided it was not the time and place to ask for the young man's name, rank, and serial number."

Sergeant Ed Cunningham, like Richardson, tugged at some gun triggers but only in a test. Covering the headquarters of General "Vinegar Joe" Stillwell in China, on December 23, 1942, he flew in the first B-24 moonlight bomber raid on Rangoon, getting the first eyewitness story of such an attack. Wondering what he could do "in case one of the gunners got shot and needed a replacement," Cunningham suggested that "I fire a few bursts from the waist window gun just to get the feel of it." So, he said, "I burned several rounds into a nearby cloud. The bursts of fire disappeared into billowy whiteness."

In reports home, the correspondents tended to treat close calls lightly. Sergeants Burtt Evans, Slim Aarons, and Burgess Scott were together on the Anzio and Nettuno beachhead, just south of Rome, when Germans fired down from the high ground along the ancient Appian Way. "One of our little annoyances," Evans wrote back, "was a 210-millimeter railway gun which popped away at us all the time I was there. After one such salute, I brushed off the dust and glass and pulled the bodies of two Canadian photographers out of the debris." Scott concurred, "Kraut observers could place a 170-millimeter shell right in the living room of any house in both of those towns."

A 500-pound bomb struck just behind Aarons, Scott, and Don Whitehead of the Associated Press. In the excitement, McCarthy mentioned in a May 9, 1944, round-robin that Whitehead, already protected by a steel helmet, managed to snatch away Aarons's plastic helmet liner. The wire reporter popped it incongruously atop his own covering, while Slim's steel fedora slipped down his nose.

Forsberg's communications command included the chain of

Stars and Stripes newspapers so, on occasion, the same soldiers contributed to both publications despite their different formulas, *Stripes* playing up last-minute local news and *YANK* seeking features that would hold up for a month or more. Andy Rooney, later the CBS TV star, was among the *Stripe*rs who contributed occasionally to *YANK*, while Sergeant Ralph Martin, the postwar author of best-selling biographies, was one who switched full-time from *Stripes* to *YANK*.

"For me," Martin said in 2000, "*YANK* wasn't just the greatest magazine of the war, it was a bonding. The guys I knew on the Paris edition and, later, all the others, became some of my best friends, hugging friends. All of us would go on to many other things, some of them even important, but nothing ever compared to this shared experience of the most exciting story in the world at the peak of our young lives."

Martin gave *YANK* a series of reports on the fighting in North Africa and, thanks to doffing his sergeant stripes and donning an officer's trench coat, had memorable encounters with Franklin Roosevelt and Winston Churchill. When the American president attended the conference in Casablanca determining North African war policy, the humble GI lunched two tables away from the commander-in-chief. Even more picaresque was the later occasion when Churchill and his generals toured the Siegfried Line at the German border. It came time for a pit stop, and Britain's doughty prime minister suggested, "Let's do it on the Siegfried Line!"

"While we were all peeing on the German tank traps in unison," Martin recalled, "the Signal Corps photographer, who wore his stripes, was ordered not to take this most symbolic picture of the entire war. I thought that if (*YANK's*) Slim or Pat Coffey were there, without their stripes, they would have taken that picture."

Rare is the correspondent who feels that copy desks back home treat his prose with proper regard, and *YANK's* writers were no different in that. Non-com Knox Burger was one of the first American soldiers to fly into conquered Japan. He cabled a story about the amazingly peaceful acceptance he met. It was, he punned, "an immaculate reception."

Knox loved the phrase, in his view the best part of his report, but an editor on 42nd Street thought otherwise. Some, for religious reasons, might take offense. A change was made to "an uneventful reception."

"Uneventful reception!" Knox still fumed at 77. "Clunk! I still haven't forgotten or forgiven." He knew who blue-penciled him and, for a while, nursed a grudge, but "we did become friends after the war."

YANK
Scoops 23

News media crave exclusives, and *YANK*'s soldiers had their share.

In the January 24, 1944, letter to the field, McCarthy exulted over a beat the London edition had, not only over its PX newsstand rival, *Life* magazine, but all the press of Britain. *YANK* revealed that Eisenhower would command the invasion of Western Europe. Fleshing out the story was Ike's picture on the front and a two-page spread of details inside.

"What made the scoop even nicer," wrote the gleeful McCarthy, "was that the issue of *Life*, which appeared in London that week, had a cover picture of General Ira Eaker as chief of the 8th Air Force. ... Eaker had already been taken from the 8th Air Force and sent to Africa."

In mid-1944, *YANK* came up with an even more dramatic exclusive. Sergeant Walter Bernstein scooped the press of the world by getting the first interview with Marshal Tito, the Communist partisan who was fighting the German occupiers of Yugoslavia. The feat caused an uproar inside high British and American military counsels, since Yugoslavia was an agreed area of British influence and out of bounds to U.S. troops. To make it worse, Allied PR officers had promised four American and British civilian correspondents and photographers that, when the time was right, they would be parachuted in for the first media encounter with the man who, as it turned out, was to become for a generation the Communist dictator of the heart of the Balkans.

Bernstein's original assignment was to Moscow, but his orders had been phrased in an opaque fashion. He was to go to Tehran and then proceed appropriately. The Iranian capital, as it turned out, was the end of that road for him, but he came up with another way to use the open-ended instructions. In Cairo, he met two Tito staffers who were there for hospitalization. They agreed to smuggle him to their leader. That involved slipping past American Army officials in Bari, Italy, trading the American uniform temporarily for Partisan garb, and walking seven nights across a chain of mountains, ducking German road patrols on the way. Tito gave *YANK*'s EM an hour. He spoke proudly of how he had managed to unite Yugoslavs of disparate religions and ethnic loyalties. He had some kind words for America, and he spoke of his needs, notably food for an underground army said to number 300,000.

So far, so good, but at that point the British liaison team at the guerrilla headquarters put the *YANK* man under effective arrest and flew him back to the American command in Algiers for appropriate punishment. Further, to take as much wind as possible out of the beat, the British hastened to fly in the four designated civilians. They outraced *YANK* into print, but at least the magazine was able to claim it was the first to hear the mountain fighter's story from his own lips.

The U.S. Army in Algeria winked an eye at Bernstein's misdeeds, and *YANK* rewarded him with a trip home. Meanwhile, *YANK* trumpeted the Bernstein story regardless of what the four civilians published. The Cairo and London editions led their issues with the account. The world edition ran three two-page accounts.

As the war wound down with the occupation of Japan, there were more scoops, some of them among the weekly's most spectacular. When Japan offered to surrender, *YANK* staffers on the

O'Brien-Craemer edition packed up for Tokyo. They had none of the required orders, but they bummed a plane ride to Guam and, from there, flew to Okinawa, where civilian war correspondents had assembled to accompany MacArthur to Japan. The *YANK* team tagged along.

At age 79, former pfc Knox Burger recalled the arrival at Tokyo's Atsugi airdrome. "When we landed we were served a meal in a large hall. There was silver cutlery and linen cloths and napkins." No mess kits for conquerors! "Later in the afternoon, evidently waiting for the most flattering light, MacArthur climbed out of his transport plane, corncob pipe in his teeth, his cap set at an old man's attempt at a jaunty angle. I watched him pose." The general gave orders for a paratroopers' victory parade.

"Ever overbearing and intrinsically superior in attitude," Burger noted, the general "weeded out all men under six feet so that the height-challenged natives would be suitably impressed."

The victory march, however, had to be delayed. Cargo ships had to unload "giant machines called prime movers—their purpose unknown to me," said the ex-pfc. They were "to participate in the parade, roaring monsters chewing up the streets and striking awe into the hearts of the Japanese."

YANK took advantage of the postponement. Both Yokohama and Tokyo were out-of-bounds for the time being, and the Japanese government had warned that it could not be responsible for the safety of troops entering the capital prematurely. There were also rumors that a Black Dragon Society might attack the incomers. Even so, Staff Sergeant Lester A. Schonberg, the Sad Sack's Baker, and five other *YANK* staffers commandeered a Dodge car and drove to Tokyo.

Schonberg was the managing editor in absentia of the *Brecksville* (Ohio) *News*, so he sent his paper a long account of what happened on that hectic first day. He described the arms the *YANK* team carried, just two .45 caliber pistols, a short bayonet attached to one man's belt, and only two clips of ammunition. In the round-trip into and out of the capital, the seven saw not one American. So far, as they were able to learn, they were the first of their countrymen to enter the Japanese heartland. A few children ran in fright as they passed, but adults ignored them or stared impassively.

Burger had written for *YANK* an eyewitness account of the first of two March 1945 fire-bombing raids on Tokyo and Nagoya. Driving into Tokyo, he was astounded to see the extent of the devastation the incendiary weapons had inflicted.

"Arguably," he said, "those horrible raids did as much to end the war as the atomic bombs."

In Tokyo, the *YANK* team hunted for a place to print edition number 21. First, they tried the two English-language newspapers, the *Tokyo Advertiser* and the *Nippon Times.* By then, the two had been folded into one after the *Times* had burned out. The combined plant had seven linotype machines with English typefaces and an old-style newspaper press that *YANK* might use. *Times* editor Togasaki received the visitors amiably and startled Schonberg with his accent.

"He speaks exactly as we do," the sergeant reported back to the readers of the *Brecksville News.* "He has just as much U.S. profanity and, with our eyes closed, we would think him one of us."

The *Times* editor explained: he was a native American. He had married a student from Japan and had followed her back to her homeland when she insisted. Their two children were American born.

Next stop was at the city's largest daily, the *Asahi Shimbun.* With Mr. Togasaki along as interpreter, the *YANK* team checked out the printing plant and put off for a day the decision about which of the two premises *YANK* would use. Back with the landing forces for the night, the *YANK* team returned the next day for a new tour of the capital, this time to find office space and lodgings. Radio Tokyo, near the past and future American embassy, was chosen for the first, and Burger recalls how he, at age 21, and 25-year-old Yeoman Robert L. Schwartz, one of *YANK*'s token sailors, took care of the housing needs. Their "patsy," he said, was "a Japanese businessman who had parked his family out of town and was living with his mistress, a former Tokyo Studios actress. They had a comfortable Western-style house on a tree-lined street not far from the *YANK* office."

"We told him that it was either two amiable young enlisted men or he'd find himself invaded by a considerably greater number of stuffy MacArthur staff officers who were likely to be less understanding in general. Also we could provide scarce foodstuffs for him as a sort of barter for space."

The actress became a dear, but temporary, friend of one of the other *YANK* staffers.

At Radio Tokyo, Burger took a desk that had been used up to a fortnight earlier by Iva Ikuko Toguri, one of the most wanted American defectors in Japan. Burger still has the four-inch-tall round box with a carved lid that he found in Iva's drawer and has kept as a dismal souvenir. Iva was said to be the notorious Tokyo Rose, the broadcaster who sought to undermine the morale of soldiers, sailors, and Marines in the Pacific by playing them American jazz and sweet-talking them into homesickness and a desire to quit. Actually, she was one of six women who took turns, but she was the only one with U.S. citizenship. Born in Louisiana, she had been a coed and a zoology major at the University of California at Los Angeles.

Starting with such leads as they could find at Burger's newly acquired desk, *YANK*'s Sergeant Dale Kramer and Corporal Jim Keeney tracked Iva down. At a conference for civilian correspondents, Keeney triumphantly presented the woman so many of them had wanted to find. It was a happy day for the *YANK* staff but the beginning of several painful years for "Rose." She was tried for treason in San Francisco. Keeney was called as a witness but contributed very little. "I was unproductive," he recollected at 91. "Since I had

The *YANK* **issue being handed out to the incoming MacArthur troops.** *(Courtesy of YANK)*

never heard any of her broadcasts I could not testify to anything she said that might have been interpreted as treason. I was dismissed after a few minutes."

Iva was sentenced to 10 years in prison at the Federal Reformatory for Women in Anderson, West Virginia, where she worked quietly as a secretary in the hospital. Released for good behavior after six years, she was pardoned by President Ford in 1977.

The finding of Tokyo Rose was a moment of excitement for the *YANK* staffers, but perhaps their greatest coup came at the end of the first week of September 1945, when MacArthur's six-foot-tall paratroopers and jangling "prime movers" paraded in for the official beginning of the occupation. The incoming troops were handed copies of a red-and-white 16-page publication labeled "*YANK, The Army Weekly*, printed in Tokyo, special surrender edition."

There were no cartoons, but the last page was a triumphant pinup, an in-your-face portrait of sultry Ella Raines in an evening dress. There was a half page Sad Sack drawn in Tokyo by Baker showing the eternal butt of GI jokes digging a foxhole to escape the "friendly fire" of recklessly celebrating buddies. Under a calculatedly underplayed headline, "Big Day on the Missouri," pages two, three, and four carried Corporal Tom Kane's photographs of the battleship surrender ceremony. Filling out the issue was copy from New York, a four-page spread about devastated Germany written by Corporal Debs Myers, and a two-page layout giving readers "a look back home since you went away," views of four cities: New Orleans, Tulsa, Evansville in Indiana, and Providence, Rhode Island. Everything had been wrapped up in the one issue, the two victories, and what the GI wanted above all else—to go home.

YANK

Send Back **24** the Story

In the middle of 1942, *YANK*'s newly uniformed civilians found hard the sudden transition from a peaceful, and sometimes effete, life to the mysterious new business of soldiering.

One question was what chance the staffer stood to be killed. Reactions to that were ambivalent. *YANK*'s first editor, Captain Spence, wavered between a stiff upper lip approach and, concerned about staff morale, a tendency to treat death as one of war's dirty little secrets. When Dave Richardson joined the staff, the captain told him, "The first thing I want you to do is to write your obituary."

Dave passed the next hour recording the college experiences that had preceded his job running copy. Waiting for more from the captain, he was fascinated by a news clipping on a desk, "*YANK*'s correspondents will go to every battlefront. ... If they live, they will send back stories of the actions in which they fought. If they are killed, other correspondents will take their place."

When Slim Aarons signed on, he heard the same, "Even lying there bleeding, don't forget to send back the story." *YANK*'s officer, said Slim, "was seeing it all like something in a movie."

The war reality for the *YANK* staffers was far from what the editor envisaged. Rather than being welcomed as fellow combatants, the staffers, although armed, often were treated like civilian reporters, just an inconvenience apt to get in the way. Diseases, not

217

bullets, caused a large share of the publication's casualties.

When *YANK* did lose a staffer, the Spence round-robin of February 2, 1943, treated it as routine business, hardly worth a mention. The letter handled the news in just a couple of lines, the rest of the report 50 times longer: "We had our first casualty last week. Red Gallagher, whom some of you undoubtedly knew, was en route overseas and cracked up. Red leaves widow, 3 kids." End of item.

Captain Basil Gallagher, to give him his full name, was one of the dozen officers *YANK* had enrolled by that time. He was on his way to oversee one of the expanding list of faraway editions. He was killed in a plane that crashed in the Brazilian jungle.

A week later, Spence reported another casualty, this one non-fatal, "(Sergeant) John Barnes is in a hospital in China recovering from effects from a crackup while on a bombing mission." This time, the editor thought the item could be put to good use prompting more front line daring, "Barnes has been on eleven (underlined) bombing raids. Ain't some of you ashamed of yourselves? Not you, Morriss, nor you, Cunningham, nor you, Burgess Scott, nor you E. White, sir!, nor you D. Richardson."

Malaria took a toll, among its victims Brodie, Greenhalgh, Morriss, and Bernstein. Burtt Evans, then in Tehran, came down with "some kind of fever." Sergeant Joe Wright developed "something wrong with his throat." Richardson finished one Southeast Asian engagement with "bandaged feet, diarrhea, and cracked lips." Pawlak was so sick in the Philippines with the mosquito-born dengue fever, "the bone breaker," that he did not utter a complaint when I gave him bad news. Hitchhiking on a truck while carrying our mail, I had dropped beneath the wheels of oncoming traffic the new telescopic lens Pawlak had requested to enhance his survival chances. When Sergeant Robert A. Ghio returned to New York after one tour, McCarthy told readers of the October 18, 1943, staff letter that "you wouldn't know the guy; he has lost about 60 pounds (to malaria and dysentery).

Pawlak earning his second Purple Heart on Okinawa. *(Courtesy of Mason Pawlak)*

Then the long-awaited close calls and Purple Hearts began to accumulate. It was a bit more like the first editor had expected. McCarthy's round-robin of March 31, 1944, shared news: "Slim Aarons was banged up but not seriously when a German shell plowed into the correspondents' house on Anzio. Aarons got national publicity on his wound because Ernie Pyle was with him at the time and mentioned him in his column which is read these days by every civilian in America."

Like *YANK*, the civilian Pyle covered the war from the point of view of the suffering soldier. Some months after his Aarons item, a sniper in the Pacific killed Pyle.

Aarons got a laugh out of the Anzio injury by writing *YANK*'s picture editor, Sergeant Leo Hofeller, "I wanted to scare (Sergeant William) Potter, (the magazine's accountant), by telling him that all my equipment was wrecked, but I was afraid he might go out and shoot himself." The happy fact was that not a roll of film was lost and not a camera scratched.

On Okinawa, cameraman Pawlak did less well. To accompany the Purple Heart he received on Angaur, he acquired a second. In the field with the Army's 96th Division, he was bracketed by shell hits. His head and face were struck with a heavy blow that reminded him of the haymakers he had met in the ring as a 16-year-old Golden Gloves boxer back in Detroit.

"It's like a thunderstorm out there," he said to medics at a schoolhouse first-aid station. "That's not water you're covered with, it's blood," he was told. His nose was broken. Shrapnel had pierced his face and torso. Then the medical shelter itself was hit. One of the medics lost a foot. With their roles reversed, the *YANK* sailor helped bandage the aid worker. Pawlak was another to get the Bronze Star. Fifty years after the war, the left side of his body still aches and his left eye never since has worked correctly.

Two other *YANK* photographers fared worse. One was Sergeant Pete Paris, a fine arts student from Syracuse University, who had been the first man hired for the staff. Editing a newspaper at Fort Belvoir, Virginia, he had heard of the plans for *YANK* and was waiting for editor Spence when General Osborn gave the officer his assignment to bring out the first issue within six weeks. Minus any other applicants, Spence employed Paris on the spot.

For the next two years, Paris was one of *YANK*'s busiest cameramen. His photos of American troops at El Guettar and Maknassy in North Africa made a cover and the first two pages of the magazine. Pictures of the first Negro unit in combat and of the French Foreign Legion followed. Then came the Sicily campaign. After nine intense months, Paris got his first home leave. When he walked into the home office, McCarthy took note that "that chubby little chin had disappeared."

With him, as a souvenir, Paris brought a huge Nazi flag, which thereafter adorned the headquarters. Sicily, Paris reported, had been a relative breeze for the American Seventh Army, "just like maneuvers." Lieutenant General Omar Bradley and the First Division's Major General Terry Allen were tops in the opinion of their troops, he added.

Not much later, in London preparing for D-day on the Continent, Paris was injured in a training jump with paratroopers, but his spirits stayed high. He wrote a cheerful letter of complaints to Annie Davis. The English food, he said,

> is excellent, as excellent as Brussels sprouts, cabbage, and a horrible ersatz concoction called sausage can be. Some of the boys are now contributing their hard earned shekels to a joint lovingly called Horsemeat Harry's when the spirit and the finance officer comes through, or the braver of us go to a place called Prunier's where you can get off pretty cheap at one pound ten including a nickel tip for a cup of coffee for the waiter.

As for lodging, Paris said that he was at "a very swanky address but that is as far as it goes." He was in a hotel annex in "a room half the size of the men's john in 205 and not unlike it, with four walls and a mirror." That glass reflector, he wrote, was a nightmare, "when I look at it from under the covers in the morning, I see this horrible creature. When I come in the door, there's another person in the room. Some day I'm going to put my foot through it."

There was savage irony in the latter comment, for two months later in the Normandy landings Paris either stepped on a land mine or was struck by artillery or a heavy weapon and a leg was torn off at the hip. A Navy medic carried him to an evacuation ship, an LST (landing ship for tanks).

The sailor filled in *YANK*'s London bureau about what happened next and a round-robin picked up the story,

> The medic said that Pete was conscious and that he got very sore when he told him he had never heard of *YANK*. He started to bawl out the medic for such ignorance, telling him in no uncertain terms what *YANK* was. The medic, of course, knew all about *YANK*. He was just trying to make Pete sore to take his mind off the leg. The last the medic saw of Pete, he was on the LST after it pulled away from shore.

The LST did not make it to England. It was bombed or shelled, drowning Pete and everyone else on board.

The other *YANK* cameraman to suffer a similar fate was Sergeant Bushemi. He was killed on Aniwetok just a few weeks after the home office had told him not to be discouraged about how difficult it was to photograph combat but rather to keep doing work that was better than he realized.

A Depression-era high school dropout and, like his father and brother, a steel mill hand in Gary, Indiana, Bushemi had learned how to use a camera and had become a staff photographer for Gary's *Post Tribune*. He won the nickname "Johnnie One Shot" for his knack of catching the critical moment in sports contests. He became known also for his daring coverage of Prohibition-era criminals.

Enlisting five months before the attack on Pearl Harbor, Bushemi joined *YANK* soon after Paris did. It was Bushemi who took the infamous aborted first cover picture of staffer Oliphant waving dollar bills as the reason GIs fight.

In the Pacific, in the view of picture editor Hofeller, Bushemi "gave *YANK* some of the greatest pictures of the war" while evincing "an enthusiasm for action photography which I have never seen equaled." As the prewar photo editor for the *New York Daily News*, Hofeller's opinion had weight.

Johnnie Bushemi, killed on Aniwetok. *(Courtesy of YANK)*

In November 1943, artist Greenhalgh, on duty in Hawaii as an editor, notified New York that Bushemi was off on a classified mission that "might get some of the war's best pictures if he is able to hold his camera above his head." The reference was to successive invasions of the Marshall and Gilbert Islands with fighting on Makin, Tarawa, Kwajalein, and Aniwetok.

In the Marshalls, covering the Fourth Marine invasion of the Roi and Namur Islands, Bushemi suffered a fracture of a finger on his left hand. It was put in splints, and he carried on with risk-taking close-up camera work, "photography at a rifle length," as accompanying writer Miller described it.

In the February 12, 1944, round-robin, McCarthy expressed relief that John was safe and sound with the Fourth Marines on Kwajalein. A letter from Bushemi had just arrived describing one bizarre episode. "During the fighting somebody discovered a dugout full of beer. About 200 Marines actually left their M1s and their BARs and ran to the place and started drinking. In order to get them the hell out of there, an officer had to rush up and break the bottles and pour the beer on the ground." Bushemi got a shot of six guzzlers.

McCarthy's relief was short-lived, for a week later Bushemi was dead. On Aniwetok Island, a knee mortar shell drove shrapnel into his left cheek, his neck, and the left leg. He bled profusely as sulfa and plasma were administered. He was carried to a Navy transport. Two hours after being hit, he died as surgeons attempted to tie off severed arteries. Miller said the photographer's last words were a request to get his photos back to 42nd Street. Spence's fantasy about proper deathbed procedure had been carried out.

Combat reactions are strange. Fellow cameraman Bill Young, on Kwajalein, heard from the Signal Corps that his colleague had lost his life. "I didn't think much about it," he admitted. "My turn was coming up. It was my first combat operation. As I looked back seven days, I should have been killed several times at least. Always, 'tomorrow is my last day.'"

YANK

6,000 25 Applicants

YANK's assignment to serve as the unique global voice of the traditionally mute men in the ranks was carried out by a haphazardly chosen staff.

Just how accidental was the process was underlined in round-robin letters midway through World War II. Six thousand, by that time, had sought a job on *YANK* while those already aboard, far from being content with their relatively good fortune, grumbled that the Army was holding its publication to just about as many privates, corporals, and sergeants as in any other company of a hundred or so members.

Put together in haste in mid-1942, the magazine had grabbed staffers wherever it could find them—some of them walk-ins, others friends of those already on the staff, still others draftee newsmen who caught the eyes of *YANK*'s officers by writing and illustrating articles about Army life for papers back home. McCarthy himself was one of the latter. A prewar sportswriter for the *Boston Post*, he found hilarious much of his chores as a corporal in North Carolina assigned to one of the Army's final mule-mounted artillery units. His column, "Postage Due," became a *Post* feature.

At first, *YANK* was a promotion-hungry soldier's paradise. During a two-week stretch in June 1942, five privates jumped up 19 ratings, the temporary managing editor Bill Richardson skipping over private first class, corporal, buck sergeant, and staff sergeant to the second highest enlisted rank, technical sergeant. All five hurdled over at least two ratings to become buck or staff sergeants. When battle-minded Corporal Dave Richardson left 42nd Street for the MacArthur theater, he pole-vaulted three ranks to *YANK*'s then highest rating, tech sergeant. With only one man for every three available ratings, *YANK* at the start was an advancement-minded EM's dream.

That soon ended, and the reverse became true. With top slots filled and circulation rising into the millions, many of *YANK*'s soldiers became more famous than most colonels, and some generals, while still stuck at the Army's humblest levels. Corporal Hargrove was a spectacular example. He was another of the newsmen who took to public print with good-natured descriptions of the ironies of soldier life and a book of his essays, *See Here, Private Hargrove*, became a best seller. Wearing two stripes when he was picked up by the magazine, he was still striving ineffectually for buck sergeant eight months after joining the staff. He was "the ranking corporal in the Army," a millionaire, as his colleagues understood it, when he sold the cinema rights to his book. When the movie came out, Hargrove took two or three of his fellow *YANK* EM for a celebration, not only at New York's fashionable Stork Club, but inside its Club Room, an inner sanctum for celebrities.

Still a corporal, Hargrove asked McCarthy to look at a clipping. A well-known journalist had joined the Marines and was asked, "Do you think you can do as much for the Marines as Hargrove has done for the Army?" As McCarthy related it in a staff letter, Hargrove told him, to "look at that, and me just a poor old beaten-down corporal!"

Hargrove's writing talents had been discovered early after his induction, so he came to *YANK* from a military public relations job. This made him the butt of some amiable joshing when McCarthy received a publicity release about the movie.

6,000 Applicants

"The MGM version of Hargrove's Army life is a lulu," McCarthy told fellow staffers on August 14, 1943.

> The picture ends with Hargrove saving somebody's life when a howitzer explodes, or something like that, and then Hargrove pleading with the authorities to release him from PRO duties so that he can go overseas with his field artillery outfit. I think I will picket the theater, when the movie opens on Broadway, with a huge placard reading: "Hargrove doesn't know how to do about-face and never slept outside a barracks in his whole Army career."

Hargrove went off to Calcutta to join Technical Sergeant Ed Cunningham, one of the pioneers on *YANK*, in putting out the China-Burma-India edition. He got an extra stripe as technician fourth grade, but not what he wanted, the other type of three-striper, an upfront plain buck sergeancy. He took to the editorship in India enthusiastically although, as he wrote back to a headquarters in October 1943, "I wouldn't wish this theater on a dog." The shakedown with Cunningham as boss had been a problem at first, with Ed a neatness freak and a cautious editor, but, since he "is a pretty nice fellow," things were turning out better. At the outset, he said, Ed had "insane fits of cleaning the room up—everything into the wastebasket and from there into the fire." It upset him. "Honest to God," it was even impossible to "keep underwear around when (Ed started) 'clearing out the trash.'" And then there was the editing. In the beginning, Ed's "over-abundance of caution" and "chief of bureau complex" had him insisting "that I show him every goddam thing I did and passed on every single comma in the book."

All that, fortunately, was past. In fact, by then, Ed had become "damned nice to work with," he was back doing the reporting he enjoyed, and he "doesn't bother me much." I think, he said, "the combination is going to work out splendidly."

Low ratings served well to keep *YANK*'s staffers in the same

From left to right: Tactical Sgt. Ed Cunningham, Joe McCarthy with his Legion of Merit decoration, George Baker of the Sad Sack, and best-seller Sgt. Marion Hargrove. Unidentified Tech Sgt. on the right. *(Courtesy of YANK)*

plight as the GIs they were to mirror, but no negative considerations were keeping thousands from trying to join the staff. For one thing, *YANK*'s professionals had the good luck to continue and even advance their peacetime careers. For another, the freedom to live out of camp and to move in and out of combat at will was a rare luxury. McCarthy and *YANK*'s officers were besieged with applications. McCarthy finished his August 25, 1943, round-robin this way: "And now will you pardon me for a minute? There is a young man outside, age nineteen, who worked for two months as assistant sports editor of the York, Pennsylvania, *Times-Chronicle* before he was drafted and wants to be a special correspondent to accompany Joe Louis on a world tour." Joe, the former world heavyweight boxing champion, was in service as a staff sergeant. McCarthy commented about the applicant, "sometimes I wish I was back with my old buddies in the 97th field artillery."

Three months later, in a letter to the field, McCarthy returned to the subject of *YANK* wannabes. "Practically everybody in the Army who knows how to type is trying to get on *YANK* and we get letters from every Congressman and general about the son of their old college roommate." One such appeal had just been made to *YANK*'s business manager, Captain Harold B. Hawley, later a major. Hawley had been circulation manager for the *Look* picture magazine, *Life*'s direct rival. "(Hawley) was approached by a man who owns several newsstands around New York. The man had a son who wanted to get on *YANK* because it was handy to a couple of his favorite bars. Captain Hawley explained about the T/O being full, and all that, but the guy went away burning. Just to get even he banned *Look* from his newsstands for the following week."

Adding to the attempted, but fruitless, rush to *YANK*, as the GI editor explained on January 24, 1944, was that "the Army isn't handing out commissions anymore and is still drafting as fast as ever." In 1942 and early 1943, he recalled, "Good writers in danger of being drafted usually got themselves commissions." As an example, he cited St. Clair McKelway, the managing editor of *The New Yorker*, who became a lieutenant colonel. "Naturally, the quality of the writers who are privates has shot upwards these last few months. Among our recent applicants are plenty of exceptionally good men from the *New York Times*, the *Saturday Evening Post*, *Colliers*, etc." One was mentioned in particular, "He is a graduate of the University of Prague and the Sorbonne and speaks all the European languages without an accent. He has worked on several European newspapers and, since he has been in this country, has sold stuff regularly to *Colliers*, *Liberty*, *The New Yorker*, *Coronet*, *This Week*, and all the big magazines. If our T/O only allowed us to pick up about 20 extra men now we could get some remarkable talent. But as things stand now, we are more or less hamstrung."

The horse was out of the barn, but McCarthy and *YANK*'s

officers decided to lock the stable with at least one change in the case of the few future additions to the staff. "We have decided that one reason why there are no ratings here to give out to guys who have worked hard for a year or more is that a lot of corporalcies and sergeancies have been swallowed up by the sergeants and corporals who came here in grade." Staff sergeants had joined as staff sergeants, and privates as privates, regardless of their particular publishing talents. "From now on if he is a sergeant he will come to *YANK* as a T5 (technician fifth class, one stripe down). If he is a corporal or pfc he will come here as a private. After all, the 6,000 or 7,000 ex-newspapermen and ex-magazine men who are breaking their necks to get on *YANK* would be only too glad to take a bust to get here."

That, said McCarthy, might make it possible to give at least a stripe to such staffers as Shehan in Italy and artist Jack Coggins on the London edition, both still privates after many months abroad.

Shehan, a sports reporter, was a prolific postwar ghost author of the "autobiographies" of golfing greats such as Ben Hogan and Sam Snead. At age 33, in 1944, Coggins was an established commercial artist who was "bugged" after the war by the fact that the *Liberty* ship to which he had been assigned for the Normandy landings got to Utah Beach only on D-day plus one. "The area (off the French coast)," Coggins remembered at 90, "was quite lively with cruisers and battleships firing at targets behind the beaches, but I had missed the big show." By then, of course, Coggins' London colleague Paris already had been dismembered and drowned in the same waters.

Coggins finally made corporal and, just before discharge, sergeant. He published 15 books after the war, including *Ships and Seamen of the American Revolution* and *Arms and Equipment of the Civil War*.

Some of the first *YANK* men were "goldbricks" in McCarthy's view, do-nothings, happy to be civilians masquerading as soldiers. One vanished for months into the land of lovely Caribbean beaches sending back not a word or a sketch for publication, nor even a letter

Joe McCarthy, flanked by George Burns and Annie Davis Weeks. (Courtesy of YANK)

suggesting he was still alive. By 1943, there had been a fairly successful effort to screen out staffers of that ilk. Even so, *YANK* remained in large part to war's end an eclectic aggregation of men of varying levels of talent and widely different backgrounds and visions of the world.

One of the fellow staffers with whom Hargrove celebrated the movie sale at the Stork Club was assistant managing editor Sergeant Justus E. Schlotshauer, whose prewar job was as advance man for the Ringling Brothers Barnum and Bailey Circus. Close to him in that calling was artist Corporal Joe Cunningham who, in civilian life, took time off each year when the circus was in town to perform with it as a clown.

Others of varied backgrounds included Sergeant Newt Oliphant, a New York songwriter, whose "Same Old Story" was a late-1930s hit. It was Oliphant who posed for the disastrous dummy cover, gleefully holding aloft the few extra dollars a month privates had been granted. He also managed to get a revealing interview with Japanese prisoners of war in the Philippines about their lives as the emperor's soldiers.

Eminent politicians, statesmen, and civic leaders were in the ancestry of some of the staffers. One such was Sergeant John Hay. Hundreds of soldiers wished to appear in print with poems both comic and sad, and it was Hay's job to select the winners for publication in the weekly's "Poets Cornered" section. John, himself a postwar lyricist on Cape Cod, was a descendent of the John Hay who was Lincoln's assistant private secretary and McKinley's and Theodore Roosevelt's secretary of state. It was the earlier John Hay who made a place for himself in U.S. history through his open-door policy for China. *YANK*'s Hay bore a striking resemblance to the picture of his forebear.

Washington bureau member Sergeant Richard Paul was the son of one of FDR's undersecretaries of the treasury. The mother of Sergeant Ecker, *YANK*'s final acting managing editor, was often in the New York headlines as a children's court judge.

At least one on *YANK* came to the magazine from his own career as a public official. Corporal Max Novack, who handled the columns "Mail Call" and "What's Your Problem," had been a New York assistant district attorney.

6,000 Applicants

Some of *YANK*'s staffers came from the far left of politics. Walter Bernstein, who got the Tito scoop, had an Aunt Sara who was one of the founders of the American Communist Party and, in the early 1930s, a secretary for the Communist International (the Comintern). Walter was one of four members of a Young Communist League cell at Dartmouth College and, though a talented postwar screenwriter for Hollywood and for television, was one of those who were blacklisted in the 1950s and found the need to write under assumed names. Like many others, he was distressed by events inside the Soviet bloc and eventually renounced Communism.

Robert Bendiner was a pfc when he joined *YANK* at age 35, after years as a political activist, including a half year on the *New Masses*, a publication which always followed the Communist line. From 1937 to 1944, he also served as editor of the *Nation* magazine. Offered a deferment as a 4F, unfit for service, he insisted on doing his share as a soldier and still speaks of his first day as a private at the Camp Upton reception station on Long Island. Assigned to KP, it was his task to put a piece of sponge cake onto mess kits already swimming in gravy, meat, and potatoes. He dumped it where he could until the mess sergeant reprimanded him. "F'chrissake," said his culinary superior, as he recalls it, "y'r ——ing up d'cake. Git it over on d'meatballs!" Another disillusioned with Stalin, Bendiner served on the editorial board of the *New York Times* from 1969 to 1977 and wrote seven books, including *The Riddle of the State Department*.

At least one *YANK* member bore the name of a hero of the American left, Private Debs Myers, a descendent of Eugene V. Debs, a spokesman for labor Socialism from 1893 until 1921. Short and stocky, "shaped like a football," as his colleagues described him, Private Myers was one of the magazine's most felicitous writers. He was the author of *YANK*'s tender tribute to FDR at the time of the commander-in-chief's death in 1945.

The personal habits of the staffers covered the range of possibilities.

While a favorite after-hours hangout of the New York staff was Tim Costello's Irish saloon with its original Thurber murals, at least two of the men were lifelong teetotalers. Private Shehan, a postwar manager of trotting racetracks, had been so impressed by the pre–World War I anti-alcoholism campaign of Father Theobald Matthew, the "apostle of temperance," that he sipped nothing but soft drinks at Costello's.

Slim Aarons, Switzerland's liberator, was another who never touched a drop while enjoying a postwar career as "the paparazzo of class," an eminent high-fashion photographer who specialized in portraits of "attractive people doing attractive things in attractive places." Never taking his cameras anywhere unless invited, dressed in white tie and tails setting off his lofty figure, the ex-sergeant melded into the most elegant scenes, gathering material for a now out-of-print fashion book, which sells for as much as $3,000 a copy when one of them can be tracked down. The legend in the fashion trade is that the shirts of the sophisticated ex-five-striper are mono-grammed in Chinese, "No starch, please."

While the magazine's first editor had the idea that soldiers are rough, tough, macho, girl-focused males, at least one of *YANK*'s most skillful editors was alien to that. After the war, as sexual mores changed, former sergeant Merle Miller published a startling article saying that from childhood he had been an active homosexual. Interviewed by Columbia University for its oral histories, fellow staffer Bendiner spoke of his friend and "the internal struggle (Miller) had carried on in the days we served on *YANK.*"

Some years after the war, says Bendiner, with

a distinguished author ... we were having lunch and Merle told us he would be having an article in the following week's (*New York Times'* magazine) about his planned emergence from his long closet life as a gay. It must have been a jolt, I imagine, to his other colleagues as it was for me. And it must have taken considerable courage to do the emerging in print in those early

234

days of the sexual revolution. Merle told me that he had first informed his mother of the article to come and that she was so dismayed by its revelation that she cut him out of her will and did not talk to him for the next six months.

By then, Miller had written seven postwar plays, plus another seven books, and was about to publish Harry Truman's oral autobiography. The latter helped repair relations with his parent.

"When he phoned her at Christmas to tell her, in passing, that he was now writing a book about President Truman," Bendiner recollects, "she said with obvious relief, 'Well, Merle, I'm certainly glad you changed the subject.'"

YANK's soldiers, of so many different origins, combined to publish an intimately sensitive record of America's role in World War II. Many went on to successful postwar careers. Ralph Martin published 21 books, including the personal stories of the Duke and Duchess of Windsor, the Kennedy dynasty, Golda Meir, Truman, publisher Cissy Paterson, Henry and Clair Luce, Prince Charles and Diana when they were together, and Winston Churchill's American mother. Weithas, who designed *YANK*, did much the same for the Elizabeth Arden cosmetics company, helping develop their special air of elegance. Hargrove wrote scripts in Hollywood until he passed away. McCarthy, the son of an operator of a students' boarding house adjacent to Harvard, left *YANK* to become an editor of *Cosmopolitan* magazine and to write books on the Kennedys and on Costello's bar.

Joe, incidentally, was no relation to Senator Joe McCarthy of Wisconsin, whose freewheeling campaign against suspected Communists in the government added the word "McCarthyism" to the dictionary. When *Look* asked him to do a profile on Edward R. Murrow, the liberal pioneer of TV journalism and one of the senator's nemeses, Joe suspected that the magazine was interested in exploiting his name. He accepted on condition that he sign himself Joseph Doyle, borrowing his mother's maiden name. The magazine agreed.

Among other Fort Bartholomew veterans who were most prolific as writers following the war was Sergeant Walter Farley, who loved horses, and, for a while, had 12 of them at his central Pennsylvania farm. He wrote 21 books for children about the Black Stallion, plus an additional 11 on other subjects. Random House, as a starter, took in $12 million on sales of them, and continues to profit in the 21st century. Some were translated into 14 languages.

Another who was among the most widely read was Sergeant Lloyd "Skip" Shearer. For 33 years, up to his death at 83 from Parkinson's disease, he filled page two of *Parade* magazine, a weekly supplement carried in 350 Sunday papers with a combined circulation of 40 million. Using the pen name Walter Scott, he answered questions about celebrities sent into him at the rate of 4,000 a week. Illustrative of the q. and a. were these:

"Did Elvis Presley wear a girdle?" No.

"Why did Jackie Kennedy favor pants?" She was bowlegged.

Sergeant Shearer's son Derek became ambassador to Finland and son-in-law Strobe Talbott was undersecretary of state, and later chief of the Brookings Institute, the liberal think tank.

Sergeant Dick Harrity had a play performed on Broadway, *Hope Is the Thing with Feathers.*

Corporal Edmund Antrobus, son of a Bond Street jeweler in London, became the chief of British travel in the United States.

Private William Saroyan, an occasional contributor of short stories, resumed a career in which he produced more than 50 books, plays, an autobiography, and collections of essays, including the 1967 *I Used to Believe I Had Forever, Now I'm Not So Sure.*

Another time-to-time contributor, Sergeant Andy Rooney, a Bronze Star recipient, became the most famous alumnus by the turn of the millennium as the curmudgeonly commentator on CBS TV's *60 Minutes.*

YANK

Owed to 26 the Brass

Managing editor McCarthy enjoyed shocking his staff by remarking that "everything we are, we owe to the brass … " He would wait a moment to let the heresy shake up his listeners, and then he would finish the sentence, "they let us alone." *YANK* had a corps of 20 or so officers whose relationship to the EM staffers was the strangest in the Army. In effect, the EM did all that mattered, deciding what to print and what to omit, while the officers mothered them.

"I spent most of my time bailing out the boys who were doing the work," commanding officer Forsberg described his function to an interviewer. "The hardest part of my job was keeping the brass off the backs of the GIs. These were a group of young military men who were acting more like civilians than soldiers. They had to scramble for everything. They got themselves into a lot of trouble."

The fact was that *YANK*, and its EM freedom, existed only because of the subtlest arrangement between the staffers and their officers. Forsberg had the power to do much as he pleased, hiring, firing, and assigning staffers, yet he deferred in great part to GI editor McCarthy. Part of the reason was that the founding instructions were that *YANK* should be the voice of the common soldier, but it was also that Forsberg's instincts as a civilian publisher were all for press freedom. Nonetheless, Forsberg's own orders were to "direct" the staff and to see to it that they behaved like good soldiers. Going

Maj. Gen. Osborn decorating Col. Frank Forsberg with the Legion of Merit award. *(Courtesy of* YANK*)*

further, the colonel's instructions made him responsible for whatever McCarthy and his crew elected to publish.

What made it work was that the Forsberg-McCarthy match was one made in heaven. The two were a mutual admiration society. Just before he suffered a fatal fall in 2002 in his Greenwich, Connecticut, home, the 95-year-old Forsberg was asked to rate staffers on a scale of one to ten. He gave only McCarthy a perfect score. McCarthy, for his part, used a round-robin when Forsberg was promoted from lieutenant colonel to full colonel to write that "We don't know anybody more deserving of a promotion."

If that appeared to be pandering to the boss, McCarthy pointed out that he already had the six stripes of master sergeant, the highest he could go as an EM. "God knows I have no more promotions coming in this war unless I want to transfer to lyric writing in the Special Service music section in return for a commission." He made the latter sound like a fate worse than being broken to private.

Forsberg was a Salt Lake City native, a Mormon, and the son of poor Swedish immigrants. He had been a teacher of economics at New York's Pace College and had used his own classroom preachments about proper business procedures to rescue the failing Street and Smith publishing company and its 54 love-story and Western magazines. The magazines carried no ads, had few subscribers, and counted on haphazard newsstand sales. The publishing boy wonder from Utah increased the number of periodicals to 56, creating *Mademoiselle* and *Charm* for an audience of young women, but those he filled with advertisements. The new cash flow saved the whole enterprise.

The Irish Catholic Bostonian McCarthy, for his part, wanted a magazine that told the often-harsh truth about the life of the GI, but, unlike some others on the staff, he had no visceral distaste for officers. He saw them as a necessity, whether or not evil. Although he enjoyed extraordinary editorial freedom, he, like the rest of the staff, was convinced without flag-waving patriotism that the war had to be fought and won. It made for an easy working arrangement with the CO.

Best-selling author Ralph Martin, at age 81, said aye to McCarthy's appraisal of *YANK*'s superior. "We must give credit to our colonel, Frank Forsberg. He was the one who gave *YANK* the fundamental shape of freedom, allowing the guys in the field to go where the war was without any chicken-shit restrictions. It was that more than anything that gave *YANK* its premier quality. You can imagine what it would have been if we had a Patton-type in charge."

Ex-Corporal Tom Shehan was of the same view.

"The colonel was a great man," he says. "I have him to thank because I was not court-martialed by Major (Robert) Strother."

Not all the officer–EM relationships below the Forsberg-McCarthy level worked so easily, and the clash to which Shehan referred was such an example. Shehan was one of McCarthy's fellow Boston Irishmen and a lifelong man of the racetracks. He was accustomed to speak his mind even as a mere two-striper. His major, on the other hand, 10 levels higher in the military pecking order, had no question about his own prerogatives. The two got into an argument one day about some matter that Shehan no longer can remember. Voices got louder. Shut up "and that's an order," the corporal was told.

Shehan persisted, so he was impressed when the major's aide, Corporal Salvatore Canizzo, a part-time printer for *YANK*, let him know on the sly that he was in hotter water than he had realized. Shehan tried to write his friend McCarthy that he needed help but the letter, as he learned, was pigeonholed at the military post office. Like all soldier mail, it needed an officer's approval marked in the envelope's upper left-hand corner. Desperate and digging himself deeper into the hole, Shehan used the major's authorization stamp to get the letter on its way. It was the formula for big trouble, and Forsberg was called upon to referee. A personality clash, he decided. Without more, he moved Shehan to another assignment. The commander's handling of the matter was typical, according to his son Eric, "He was always a peacemaker, even between me and my brother, Lars."

Walter Bernstein agreed about Forsberg's lenient treatment of the magazine's EM staff, although, based on his experience in the

YANK's final cover. (Courtesy of YANK)

ETO, he saw that area's commanding general, Eisenhower, as the agent through whom *YANK*'s liberty came in Europe. Reminiscing in 2001, the ex-sergeant put it this way:

> There are all sorts of generals, those who are smart—not enough—and those who are stupid—more than enough, those who know enlisted men and those who don't. Eisenhower was one who did. Not from having been one—he was denied that privilege—but because he had a gift for sympathy, and this filtered down. *YANK* was uncommonly free. Its writers wrote for their peers. They truly represented their peers. That they were allowed to do so came first from the officers immediately in charge of *YANK* who protected and fought for this right but, ultimately, it had to come from on high. What filtered down from Eisenhower was respect for that right.

In Forsberg's view, his officers and the EM of the weekly got along well, but the Naples rumpus was not unique. The gingerly relationship occasionally broke down. When one officer was transferred out of *YANK*, an EM staffer exulted in a letter to a confidant in the home office, "Jesus, there was a first rate bastard for you. I don't think his own mother could trust him, and he had the personality of a third rate pimp. Far be it from me to malign the dead but I hope the sonovabitch winds up in a graves registration unit for the duration plus."

In *YANK*'s early days, when an officer gave an editorial instruction to Sergeant Hofeller, McCarthy's main assistant, Shehan, recalls what happened. Hofeller asked Forsberg to send him back to his original outfit at Fort Knox. Why, he was asked. Because, he said, if I do what I have been told to do it will be such a mistake that I will be shipped back there anyway. The directive was withdrawn and Hofeller stayed.

Officers have a mystique in EM eyes until events modify it, and *YANK* was no different in that. Both Saipan's Major Craemer and decorated Sergeant Georg Meyers still speak of an incident between them that altered perceptions. Meyers's one-man beat early in the war was Alaska and the bleak Aleutian archipelago pointing south toward Japan. He was alone in the area until young Craemer, then a captain, was, to use Meyers's words, "dispatched from New York to supervise his detachment, me."

The two were in the battle for the Aleutian Island of Attu, where 500 Americans and all but 11 troops of the 2,649 Japanese garrison were killed. The officer and his one-man command bedded down together in a pup tent and a double sleeping bag and slept through a suicidal banzai charge.

"In the intervening six decades," says Meyers, "I have often reminded Captain Craemer of my discovery that night that it was a source of great comfort to learn that officers smell as bad as enlisted men." Under the frozen battlefield conditions, Craemer had gone three weeks without changing his long john underwear and his heavy wool socks. As he mentions now, he thought it was "a bit uncalled for" when his sergeant announced "loud and clear" in the frost of dawn that "he hadn't previously known that an officer could smell so bad."

Then, Craemer said, after the victory was won and he boarded a transport, he found "what Georg had been up against. Inside the ship, I hastened to remove my clothes, which suddenly began to stink. I quickly headed for a shower, tossing all my clothes overboard without breaking a stride."

The situation of *YANK*'s officers was anomalous, the EM getting the bylines and the fame, while they settled for such ephemeral benefits as officer clubs and hard liquor allowances. Some, accustomed to peacetime editing, found it hard to keep hands off the magazine contents. There was one example when Major Craemer on Saipan received a radiogram from fellow major Josua Eppinger in Hawaii, telling him to kill a certain "Mail Call" complaint about officer privileges.

Eppinger was the forceful former city editor of the *San Francisco Examiner* and had just taken over for *YANK* in Honolulu. It was his first experience in charge of a publishing operation over which he was supposed to have no editorial control.

Craemer was astonished. How could he comply when EM were considered to be in charge of the magazine contents? Not only was

YANK's first veteran reunion, in early 1946. *(Courtesy of YANK)*

Corporal O'Brien insistent on his prerogatives as editor, but "Mail Call" and its contents came from the world headquarters where other EM, using their sovereign authority, had signed off on it.

Making it worse, Eppinger had not sent the message as a friendly suggestion from a fellow officer but, in keeping with military custom, had labeled it as an order from COMGENPOA (the commanding general of the Pacific Ocean Area) addressed to the commanding general of base command headquarters on Saipan, Craemer's local superior. Going to friends there, the officer appealed for a Saipan command "respectful" request for COMGENPOA to change its mind. The "Mail Call" item already was in print, he said. It would cost a small fortune to rerun the issue. The Saipan leadership agreed, and was amazed by the Honolulu reply. COMGENPOA could not find any such order in the general's files. The matter went dead. Craemer relaxed, and then made a discovery. Though his rattled plea had been made in good faith, it was a falsehood. The film from New York with the offending "Mail Call" item had not even arrived. There was no press run to scrap. When the film did get to Saipan, it was run off intact.

Craemer cites another odd example of the strange officer–EM roles on *YANK*. After an issue ran off on his presses, it had to be distributed across the wide spaces of the Western Pacific. Again, it was out of the hands of the officer in charge. Craemer recalls that the mysterious GI who handled that came and went as he pleased, rarely coming back without another huge order. "He didn't ask me for help or advice. That could mean that he required none or that he had no expectations that I was capable of helping."

The strange EM's name is none the former officer can recall, but ex-Sergeant St. George was sure it was Sergeant Walter Maxwell:

He didn't look like a soldier. Many higher ranks, on meeting Walter, assumed he was a civilian. Unless pressed, he did not admit he was a soldier. But he was a magician when it came to moving magazines.

Some *YANK* officers accepted in good grace their basically subordinate role. As editor of the Calcutta edition, Hargrove, in October 1943, found it hard to believe his good fortune when Lieutenant Colonel Donald B. Thurman was borrowed for a while to serve as the magazine's officer in charge for the Burma, India, China theater. Hargrove wrote about it to Annie Davis Weeks. "This Thurman is unbelievable, an all-round swell egg. He never bothers my work and he never throws his weight around. But, whenever we need anything, he goes to bat for us just like that. It's a pity he's decided to rejoin the Army." The officer was returning to his regular outfit.

Thurman, Forsberg, Weeks, and Craemer were by no means the only *YANK* officers who worked selflessly out of the light of publicity to help the well-advertised EM succeed at their jobs. Major Harold B. Hawley in the MacArthur theater was among many such individuals. In a letter dated December 12, 1944, he filled in Annie at the home office on how things were going Down Under, "I have never put in such a tough year at any time of my life, trying to get out a decent publication." From the general's staff, he said, he was getting "constant sniping, sharp shooting, and bickering. I would rather drive a milk wagon than hold down such a job in civilian life at any price. If this is the Army, I've had my bellyful." Were *YANK* to submit peacefully to the local demands, he said, "others (will be) sure to follow and our fine worldwide publication will degenerate into a series of theater poop sheets."

Three months later, Hawley had more to say, "My morale is fighting for room with the tongue of my shoe. I am beating my brains out here trying to keep these people from stealing the baby. ... It's exactly where it was fifteen months ago when I came here. No sane man would take (this job)."

Finally, on June 2, 1945, Hawley had good news. After 18 months "in this hornets' nest," he told Annie, the MacArthur staff finally had agreed to "keep their dirty hands off our baby." The news, he said, "that MacArthur holds the reins in the entire Pacific area might have been bad if the fight wasn't settled, but now it's okay and we can devote ourselves to publishing a magazine." Perhaps it was better late than never, but, in another 10 weeks the war was over, and, less than half a year after that, *YANK* was no more.

The strange alliance of *YANK*'s EM and officers had given the millions of enlisted men and women a voice unique in the story of warfare. Troop morale had been fostered. *YANK* held a share in the victory.

Afterword

YANK had no precedent and is unlikely ever to be duplicated. Enlisted men were treated as a special category inside the Army with an autonomous voice. The officer corps had no similar mouthpiece.

Created to cope with expected morale problems in a vast citizen army, it was recognized from the start as an experiment that was revolutionary and risky. The statement from President Franklin D. Roosevelt published in the first issue proved to have been more than just ghostwriter rhetoric:

> *YANK* ... cannot be understood by our enemies. It is inconceivable that a soldier should be allowed to express his own thoughts, his ideas, and his opinions. It is inconceivable that any soldier, or any citizen for that matter, should have any thoughts other than those dictated by their leaders.

By contrast, at the very time *YANK* published the commander's words, farmhouses across the countryside of one Axis enemy, Italy, were covered by huge scrawlings still visible after the war, *Mussolini ha sempre ragione* (Mussolini is always right).

That FDR's offer of GI editorial freedom was meant to be taken at face value was reflected at the conflict's end in War Department circular 292 of September 25, 1945, directing the magazine to fold up by the last day of the year. Initialed by Chief of Staff Marshall in the name of the secretary of war, the instruction said that the time had come to mute "the official voice of the enlisted man," since the shooting had stopped and "material suitable for the mission of *YANK* no longer exists." Less extraordinary traditional journalism would take over.

YANK's relationship to the "gentlemen" of the officer corps had been peculiar. Where it most counts, the latter had carried the burden of leadership and often, giving courageous example, suffered a higher death toll. *YANK* implicitly recognized that on September 29, 1943, when it listed those who had thus far died. Fifty-eight thousand enlisted personnel and 16,000 officers had lost their lives, more than one officer for every four GIs despite the far more numerous ranks of the enlisted.

Some of *YANK*'s EM, such as the Sad Sack's Sergeant Baker, became better known than many a general, and various lower ranking officers envied GI access to the magazine's pages. Shavetail Jerome Snyder, a second lieutenant, wanted to be a cartoonist, but each of his offerings, however clever, was rejected. "Please, *YANK*," he wrote, "let us in. God knows the officers need it. Don't fear destruction of the uncorrupted enlisted character of your publication." No, was Corporal Novack's answer. "*YANK* feels that the GI point of view is distinct from that of the commissioned personnel. Enlisted men make up *YANK*'s staff entirely and enlisted men alone contribute to the pages."

Major Harry S. Ashmore, at Indian Gap, Pennsylvania, a career journalist, seconded the "Mail Call" editor on that. "That the GI point of view differs from that of the officers makes sense," he wrote, adding that "you speak eloquently for the GI, a converted civilian whose tough, bawdy, sentimental spirit enables him to survive without loss of dignity." The key to the phenomenon that was *YANK*, according to Brigadier General Harold W. Nelson, the chief of military history, was that "a few perceptive men realized that our soldiers would be civilian in attitude." The GIs would be patriotic, they felt sure, "but they would be realists. They would be literate, outspoken, and accustomed to weekly magazines as sources of information, intellectual nourishment, and relaxation." *YANK*, according to Nelson, filled that bill.

From the point of view of editor McCarthy, the soldier publication met its challenge, especially in its mature years, by simple honesty, at least the truth as soldiers saw it. In the 67th and final round-robin, on November 15, 1945, he told the rapidly dissolving staff, "I don't like to get sentimental about things like this but I'd like you to know that it was very good to work with you. ... I don't know any enlisted man who stayed with us whom I would put down in my book as a punk or a bastard, and that is a pretty remarkable thing to say about an organization of more than 150 temperamental writers, photographers, and artists. And that goes, too, for most of the *YANK* officers. You know as well as I do the *YANK* officers that it doesn't go for, and there were very few of them, all things considered."

The magazine, the departing editor said, had been put together with uncommon "smoothness of feeling and good nature and so little pettiness and phoniness."

It was honest, he said, and that gave it a "greatness that very few publications possess," a particular form of eminence that "can be best described by resorting to a very distasteful and coarse but explicit colloquialism. *YANK* was great because, unlike other wartime magazines and newspapers ... and unlike most other branches of the armed services, it had practically no horse shit. ... I know we had the best jobs in the Army."

YANK closed down on New Year's Eve 1945 with extraordinary accolades. General Eisenhower, by then successor of Marshall as chief of staff, gave the magazine, like any other departing GI, an "honorable discharge" certificate. It ran as the final *YANK* cover. It was awarded, it read, "as a testimonial of honest and faithful service to this country."

In giving commanding officer Forsberg the high honor of the Legion of Merit, an honor bestowed also on McCarthy, morale supervisor Osborn described the vanishing publication as "the most widely read and beloved magazine in the history of the Army." All members of the staff received the Army Commendation Medal.

Especially pleasant was a salute a year earlier from the magazine of the Marine Corps, the *Leatherneck*, whose initial comments on *YANK* were summed up by Forsberg as a "sissy" publication. By mid-1944, the *Leatherneck* had changed its tune. It wrote:

> Shortly after *YANK* ... began operations, the editors of *YANK* and the *Leatherneck* became involved in a verbal barrage. ... The reason for our somewhat critical (commentary) was their chesty attitude. ... It was sort of juvenile ... (Our view) was a little like the neighborhood bully's street corner meeting with the new boy on the block. We were speaking ... with twenty-six years of Marine publishing behind us and were somewhat crusty. Since then we have scrubbed some of the barnacles off us ... *YANK* is a true mirror of the average soldier ... a real honest-to-God soldiers' paper. Glad to have you aboard, *YANK!*

YANK's circulation exceeded 2.5 million in 41 countries. Readership, in Forsberg's calculation, topped 4 million, as copies passed from hand to hand. *YANK* was credited with being the first global periodical, paving the way for civilian publications that followed. As the staff broke up, Forsberg became a much-decorated ambassador to Sweden, the land of his forebears. Several staffers, including Hargrove and Bernstein, went to Hollywood as script writers. Others wrote best sellers or became award-winning cameramen and artists.

In the final issue on December 28, 1945, *YANK*'s EM had a few suggestions for the Army. How about abolishing the wall between EM and officers, just having "a promotion ladder going straight from private to general?" If not that, why not have similar quarters, food, facilities, and uniforms for all in service, except for insignia indicating rank? Unsigned, the farewell *YANK* comment described what it considered "the most distressing spectacle of this war and the most disgusting for some soldiers who had a slight pride in the fact that they were said to belong to a 'democratic army.'" Those were the signs reading "Off limits to enlisted men." "The idea that the technical artificiality of rank," the article insisted, "useful only to clarify the chain of command, could entitle one man to eat in a good hotel and banish another to a fly-specked zinc counter has no place in any army that represents the United States."

Rank and extra responsibilities merit privileges, so not all the parting GI advice became the law of the land. The same was true of an American Legion post a few staffers organized after the demobilization. Designed to bring some liberal ideas to an organization most of them saw as hidebound, the post was named for Pete Paris of *YANK* and Greg Duncan of *Stars and Stripes*, both of them EM killed at the battlefront. The celebrated Hargrove was tapped for post commander. The Legion tried it on for a while and then expelled the post as too far left politically. With that the group dissolved. The war was over and *YANK*'s unique mission was at an end.

The consensus among *YANK* veterans is that the publication was one of a kind, never to be repeated. The reason is that never again is there likely to be "quite the citizen army we had in World War II," says Dave Richardson, who relished his new simulated rank of lieutenant colonel when he joined *Time-Life* as a war correspondent. Ralph Martin concurred. "It's hard to imagine that there can be another *YANK* or any other soldier publication getting that kind of freedom and keeping that kind of quality." *YANK* linked soldiers on battlefields a world apart from one another, but television does that now, "so the need for another *YANK* is doubtful," is the opinion of artist Coggins. "*YANK* did a great job at the time," he says. A global cataclysm made *YANK* necessary. "Let's hope there is no repetition."

These days, different wars, such as the struggle against faceless terrorism, will always pose the problem of maintaining morale. In a defense against terrorism, not just troops abroad but the population at home will need help in sustaining courage and the will to endure. The challenge now will be to find some other way in a changed world situation to accomplish abroad, and at home, what *YANK* did for the troops in World War II.

Author's Note

An assignment I had each year in the late 1930s as a *New York Herald Tribune* cub reporter was to write a story on the eve of Memorial Day saying that there would be the usual holiday veterans' parade up Riverside Drive to the tomb of General Ulysses Grant. To brighten the copy, I would track down the oldest marcher. He was always a Civil War veteran of 92 or 93 years of age and, uniformly, had been so young that he had never understood what went on in the battle between North and South, and thus had little to relate.

Now that almost as much time has elapsed since World War II as divided us then from the Civil War, I welcomed the suggestion of *YANK*'s old commander that I write this history of our magazine. It was an opportunity to record a little more about what went on in the minds of World War II soldiers. Now, at 90 myself, I wanted to recount more of our global war than those wearers of the blue of 139 years ago were able to tell about their experience of war.

Bibliography

YANK was banned from competing for civilian readership inside the United States during the war, but Arno Press of New York, in 1967, published in four thick volumes all 150 of the master editions.

Several selections of *YANK* articles, photographs, cartoons, and sketches have been published. They include three collections of *YANK* editorial materials selected by former staffers: *The Best from YANK, The Army Weekly*, E. P. Dutton & Co., New York, 1945; *YANK, The GI Story of the War*, Duell, Sloan and Pierce, New York, 1947; and *YANK, The Story of World War II as Written by the Soldiers*, by the editors of *YANK, The Army Weekly*, Greenwich House, New York, 1984. A fourth collection, *YANK, the Army Weekly*, was published by St. Martin's Press in 1991, with Steve Kluger as editor. A compilation of reminiscence essays by 44 former staffers, entitled *Close to Glory*, edited by Art Weithas, was published by the Eakin Press of Austin, Texas, also in 1991.

Among many other books by or about *YANK* staffers are Sergeant Ralph Stein's *What Am I Laughing At?*, Whittlesey House, New York, 1944; a collection of his comic cartoons for *YANK, Mack Morriss, South Pacific Diary, 1942–1943*, edited by Ronnie Day, the University Press of Kentucky, 1996; and *Inside Out: A Memoir of the Blacklist*, by Walter Bernstein, Alfred A. Knopf, 1996.

About the Author

Barrett McGurn served 14 months in 1943 and 1944 as a *YANK* combat correspondent in the Western Pacific, finishing service in 1945 as the magazine's national bureau chief in Washington, D.C. He was a four-stripe staff sergeant. After the armistice, he returned to the *New York* and *Paris Herald Tribunes,* serving for 15 years as a foreign correspondent based in Rome, Paris, and Moscow. From both Long Island University, in 1956, and the Overseas Press Club of America, in 1957, he received awards as the year's best American overseas reporter for eyewitness accounts of the mid-century insurrections in Algeria, Tunisia, and Morocco, as well as the 1956 Hungarian Revolution. As a foreign service officer from 1966 to 1972 in Rome, Saigon, and the State Department, he served for a year as director of the American government center for the news media in Vietnam. In 1972, Secretary of State William Rogers presented him with the State Department's Meritorious Honor Award. He has written five other books: *Decade in Europe* (E. P. Dutton, 1958); *A Reporter Looks at the Vatican* (Edward-McCann, 1962); *A Reporter Looks at American Catholicism* (Hawthorn, 1966); *America's Court: The Supreme Court and the People* (Fulcrum Publishing, 1997); and *The Pilgrim's Guide to Rome for the Millenium* (Penguin/Viking, 1999).

American History Books from Fulcrum Publishing

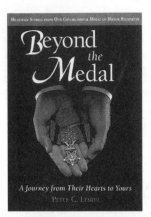

Beyond the Medal

A Journey from Their Hearts
to Yours

Peter C. Lemon

ISBN 1-55591-358-X
7 x 10 ✩ 224 pages ✩ HC $29.95

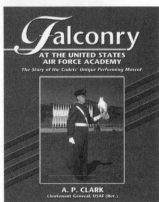

Falconry
At the United States
Air Force Academy
The Story of the Cadets'
Unique Performing Mascot

A. P. Clark

HC $34.95 ✩ ISBN 1-55591-497-7
PB $17.95 ✩ ISBN 1-55591-487-X
8¹/₂ x 11 ✩ 96 pages

Wit and Wisdom
of Politics
Third Edition

Chuck Henning

ISBN 1-55591-333-4
5¹/₂ x 8¹/₂ ✩ 288 pages ✩ PB $14.95

Fulcrum Publishing
16100 Table Mountain Parkway, Suite 300, Golden, CO 80403
To order call 800-992-2908 or visit www.fulcrum-books.com
Also available at your local bookstore.